TOP-DOWN DESIGN OF HIGH-PERFORMANCE SIGMA-DELTA MODULATORS

THE KLUWER INTERNATIONAL SERIES
IN ENGINEERING AND COMPUTER SCIENCE

ANALOG CIRCUITS AND SIGNAL PROCESSING
Consulting Editor: **Mohammed Ismail**. *Ohio State University*

Related Titles:

TOP-DOWN DESIGN OF HIGH-PERFORMANCE SIGMA-DELTA MODULATORS

by

Fernando Medeiro
Universidad de Sevilla

Angel Pérez-Verdú
Universidad de Sevilla

and

Angel Rodríguez-Vázquez
Universidad de Sevilla

KLUWER ACADEMIC PUBLISHERS
BOSTON / DORDRECHT / LONDON

A C.I.P. Catalogue record for this book is available from the Library of Congress.

ISBN 978-1-4419-5067-3

Published by Kluwer Academic Publishers,
P.O. Box 17, 3300 AA Dordrecht, The Netherlands.

Sold and distributed in North, Central and South America
by Kluwer Academic Publishers,
101 Philip Drive, Norwell, MA 02061, U.S.A.

In all other countries, sold and distributed
by Kluwer Academic Publishers,
P.O. Box 322, 3300 AH Dordrecht, The Netherlands.

Printed on acid-free paper

A Mª Ángeles

A Alberto y Aitana

A Nuestros Padres

Contents

List of Figures

List of Tables

Preface

The interest for $\Sigma\Delta$ modulation-based A/D converters has significantly increased in the last years. The reason for that is twofold. On the one hand, unlike other converters that need accurate building blocks to obtain high resolution, $\Sigma\Delta$ converters show low sensitivity to the imperfections of their building blocks. This is achieved through extensive use of digital signal processing – a desirable feature regarding the implementation of A/D interfaces in mainstream CMOS technologies which are better suited for implementing fast, dense, digital circuits than accurate analog circuits. On the other hand, the number of applications with industrial interest has also grown. In fact, starting from the earliest in the audio band, today we can find $\Sigma\Delta$ converters in a large variety of A/D interfaces, ranging from instrumentation to communications.

These advances have been supported by a number of research works that have lead to a considerably large amount of published papers and books covering different sub-topics: from purely theoretical aspects to architecture and circuit optimization. However, so much material is often difficultly digested by those unexperienced designers who have been committed to developing a $\Sigma\Delta$ converter, mainly because there is a lack of methodology. In our view, a clear methodology is necessary in $\Sigma\Delta$ modulator design because all related tasks are rather hard. For instance, since $\Sigma\Delta$ modulators are intrinsically nonlinear circuits, their detailed analysis and modeling is cumbersome; this is often overcome through the use of rather simplistic models, thus imposing the necessity of simulation for fine-tuning the design. Simulation itself is also a challenge because flat electrical simulation – the usual strategy for analog design – is useless for estimating the overall performance of $\Sigma\Delta$ modulators. Not to mention the difficulty of measuring high-performance A/D converters.

This book is intended to palliate these problems. In addition to the basic knowledge needed to design and characterize $\Sigma\Delta$ modulators, this book offers a systematic methodology, supported by dedicated CAD tools, which allows designers to achieve a given set of modulator specifications in very short design cycles. This methodology is supported by an extensive analysis and modeling of the mechanisms which degrade the modulator performance, and is demonstrated through the development of state-of-the-art CMOS prototypes.

The book is intended for a large audience: from engineers with a background on analog circuits but no previous experience on $\Sigma\Delta$ modulator design, to the experienced $\Sigma\Delta$ modulator designers. On the one hand, those engineers who have never designed a $\Sigma\Delta$ modulator will acquire the necessary insight on how $\Sigma\Delta$ modulators work, how the error mechanisms which degrade their performance can be controlled, what kind of tools can be applied, etc. On the other hand, those $\Sigma\Delta$ designers who, with great effort, have just developed an integrated prototype whose resolution is several bits smaller than expected, will find strategies to improve their designs. Finally and mainly, this book is intended for those designers who, having already developed a working $\Sigma\Delta$ modulator, want to improve their productivity through the incorporation of design methodologies to optimize the design and to shorten the design cycle.

The methodologies presented in the book have been implemented in several software tools that the interested reader can get through the following E-mail address: medeiro@imse.cnm.es

Seville, Spain

F. Medeiro
B. Pérez-Verdú
A. Rodríguez-Vázquez

Acknowledgments

The authors wish to thank Rafael Domínguez-Castro for his valuable inputs to the content of Chapter 5 and Juan Domingo Martín for his help in the development of the software.

This work has been partially supported by the CEE ESPRIT Program in the framework of the Projects #5056 (AD2000) and #8795 (AMFIS), and by the Spanish CICYT under contract TIC 97-0580.

Chapter **1**

Introduction

Modern electronics systems in computers, communications, automotive, instrumentation, etc., are mostly mixed-signal systems. That means that some signals (inputs, outputs or internal signals) are analog, while other are digital. In some applications, the digital circuitry operates as a control, calibration or storage system, while the processing is made in the analog domain. Such approximation might be suitable where the minimization of the power consumption and occupation area have priority over the resolution [Nish93]; as, for instance, in microsystems (sensors + analog circuitry + digital circuitry) used for acquisition and on-line processing of images [Espe96]. However, in many applications only the inputs and outputs are analog; the use of analog circuitry is, thus, restricted to the interfaces (signal conditioning and analog-to-digital or digital-to-analog conversion), while the processing is made by digital circuits (DSP) [Gray87]. This has promoted the use of digital techniques in applications traditionally reserved to the analog ones. Such trends are based on the simplicity of the design and testing, robustness, flexibility and easy programming of digital circuits with respect to the analog circuits.

The conventional technique for implementing mixed-signal electronic systems is to interconnect components or sub-systems on a board. They can be selected from available integrated circuits or designed ad-hoc as Application Specific Integrated Circuits (ASIC). The added value of these *systems-on-boards* is basically due to the quality and novelty of its function, as well as to its usefulness for industrial applications. In many cases, the handicaps of this realization technique (dependence on the market, large dimensions and power consumption, etc.) do not reduce the added value, for instance, in a big plant. Nevertheless, in massive production applications (communications, automotive, computers, environment, etc.) or in miniature and/or low-power systems (portable electronics, bio-implants, etc.), just implementing some function is not enough. This being the case, *integrating* the complete system on a *monolithic substrate* (*systems-on-chips*) may be critical for its success. In addition, this technique increases the robustness of the circuitry. Forecasts for the beginning of the next century tell us that 45% of total sold electronics systems will be mixed-signal ASICs, which justifies the development of the knowledge and expertise needed for their implementation.

Furthermore, mixed-signal ASICs offer interesting challenges to the traditional analog circuit techniques, from purely architectural aspects to the development of specific CAD methodologies and the design of basic cells compatible with digital technologies.

- Firstly, the increasing power consumption of DSP circuits, together with the market requirements, makes it necessary to develop faster, more precise analog interfaces, with reduced power consumption, which implies that classical architectures must be renewed.

- Secondly, analog circuits must be implemented in VLSI CMOS technologies optimized for digital circuits. Such technologies have low-voltage supply, short channel MOS devices but with excessive threshold voltage, poor matching properties and, often, deficiently modelled for analog usage [Sans96][Tsiv96].

- In addition, proximity of noisy digital circuits disturbs the analog processing, which can result in a completely wrong functionality. Appropriate restrictions should be adopted in all design phases to avoid such degradation [Vitt85][Verg95].

- Finally, the parallel design of the analog and digital parts of a mixed-signal ASIC requires shortening the design cycle of the former, which implies the use of analog, mixed-signal CAD tools. Unfortunately, the degree of development and diffusion of such tools is far from that of the digital CAD.

In this scenario, the works in this book are intended, first, to develop architectures and design techniques for high-performance CMOS A/D converters, with special emphasis on the minimization of the power consumption; and, second, to develop methodologies that allow to transfer and re-use the knowledge by mixed-signal ASIC designers with little experience in analog circuits.

The basic specifications of an A/D converter are: resolution (given by the full-scale dynamic range in effective number of bits), and speed (given by the maximum signal frequency, f, that can be processed with no degradation of the resolution). Both figures are correlated by the conversion technique (incremental, successive approximations, algorithmic, interpolative, flash, etc.), by the circuit imperfections (parasitics, transition frequency, mismatching, etc.) and, finally, by thermal noise [ADev86][Plas94]. The latter limits the maximum resolution *efficiently* achievable at each conversion speed. This limit can be approximated by a negative-slope straight line in the *Resolution - log(f)* plane [Nish93], and goes to higher resolution at a rate of 2dB/year (that is, 0.33bit/year) [Swan95]. Exclusively specifications below that line are efficiently implementable and only if proper conversion techniques are used. Because of that, it is interesting to divide the acceptability region into several sub-regions, corresponding to the converter architecture which is more suitable to achieve given resolution-speed specifications. Fig. 1.1 shows a possible division based on the results of A/D converters found in literature and on the analysis of basic properties of each architecture [ADev86]

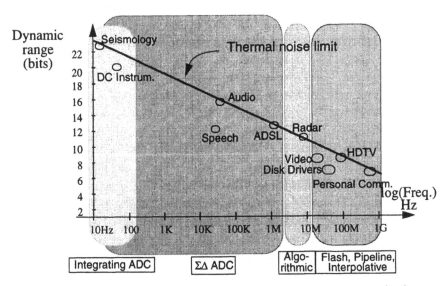

Figure 1.1: Aplication range of several modulator architectures in the resolution-speed plane. The thermal noise limit position corresponds to state-of-the-art CMOS data converters reported at the *IEEE International Solid-State Circuits Conference* (ISSCC'96) [Good96] and in the *IEEE Journal of Solid-State Circuits.*

[Plas94][Raza95][Nors96]. For a better understanding, some significant specifications have been drawn. As can be seen, the so-called ΣΔ converters (also noise-shaping converters) cover a wide region of the resolution-speed plane, including the highest resolution zone [Cand92][Nors96]. Thus, these converters are suitable for a large amount of interface applications in mixed-signal ASICs.

ΣΔ converters, introduced by Inose et al. in 1962 [Inos62], operate with redundant temporal data, obtained using *oversampling* with low-resolution quantizers (one-bit quantizers in many cases), and apply signal processing techniques (averaging in the simplest case) to combine these temporal data, thus increasing the effective resolution. Somehow, this temporal parallelism is the basis of the robustness of the ΣΔ converters, which, starting from the earliest implementations (in the audio band [Cand85]), have widened their application range from instrumentation [Sign90][Leme93] to communications [Bran91b][Bair96][Op'T93][Yin94], as shown in Fig. 1.1. Unlike traditional converters, which need high-precision building blocks (alternatively, correction mechanisms) to get global precision, ΣΔ converters show very low sensitivity to the imperfections of the circuitry, at the price of extensive use of digital processing. Therefore, these architectures are adequate to achieve high-resolution A/D conversion in VSLI technologies, more suitable for implementing fast digital circuits than accurate analog circuits.

These advantages have prompted the interest of the researchers in the oversampling converters, so since the middle eighties a large amount of works related to these converters have been published in the more prestigious international forums. As a consequence of this research activity, there are a lot of architectures available to implement ΣΔ converters, from simple single-loop low-order [Bose88][Cand85][Inos62] to the more sophisticated with multiple feedback loops, cascade connection or multi-bit quantization [Bahe92][Bran91b][Carl89][Chao90][Haya86][Lee87][Leun92][Op'T91] [Will91][Yin94]. Moreover, most of these architectures have been successfully implemented, thus CMOS ΣΔ converters with 20-bit effective resolution in instrumentation [Leop91][Sign90][Nys96], 16-bit in audio and data acquisition [Bran91a][Ferg91][Toum92][Yin93], and up to more than 12-bit [Bran91b][ZCha95] in communications are feasible.

However, two factors have restrained the use of these converters in industrial ASICs: on the one hand, the above mentioned proliferation of modulator architectures, with miscellaneous criteria regarding the choice of their design parameters, is difficultly digested by inexperienced designers. On the other, the meagre presence of specific tools for synthesis and verification, renders the design optimization a very resource-consuming, slow process, especially when, as usual, the dominant error source is not the quantization noise but those intimately related to the analog circuitry imperfections [Haus86][Yuka87]. Because of those two factors, a high specialization of circuit designers is required, which is not always possible on the industry side.

In this sense, this book is aimed to facilitate the approximation of non-experienced designers to the topic, through the introduction of a CAD methodology that enables the automation of the synthesis tasks in the design of switched-capacitor (SC) $\Sigma\Delta$ modulators, as well as their verification through behavioral simulation.

The content of this book is organized as follows: In Chapter 2 some general aspects of the oversampling converters and the functioning principle of the basic architectures are briefly introduced. In addition, present trends of the architectural design are summarized and the state of the art of A/D converter design is revised. In Chapter 3 the error mechanisms that affect the performance of SC $\Sigma\Delta$ modulators are analyzed and accurate models, taking into account a large amount of non-ideal phenomena, are proposed. By using the results of such analysis, behavioral models are presented in Chapter 4. These describe, as analytical expressions or computation algorithms, the behavior of the building blocks of the modulator. Such models are incorporated into a simulation tool called ASIDES that performs time-domain behavioral simulations of arbitrary modulator architectures, including many non-ideal effects. Chapter 5 is intended to describe two tools (SDOPT + FRIDGE) whose vertical integration assists the automatic design of SC $\Sigma\Delta$ modulators. These tools support the sizing tasks, both at the modulator level (that is, finding the terminal specifications for the building blocks, as a function of the modulator specifications), as well as at the cell level (that is, finding the transistor and passive component sizes to fulfil the cell specifications).

The effectiveness of the CAD methodology is shown through three IC implementations in CMOS technologies:

a) A second-order $\Sigma\Delta$ modulator, described in Chapter 5, designed to be included in an energy metering ASIC front-end. Experimental results show that the prototype fabricated in a 0.7µm CMOS technology has 16.4-bit effective resolution at 9.6kS/s. The power consumption of the modulator is only 1.71mW (5-V supply) operating at 2.5-MHz clock rate.

b) A fourth-order $\Sigma\Delta$ modulator, described in Chapter 6. The specifications, 17bit at 40kS/s, were fulfilled with a cascade 2-2 architecture. Measurements on a 1.2µm CMOS prototype show an effective resolution of 16.7bit at 40S/s with a power dissipation of 10mW (5-V supply and 5.12-Mhz clock rate).

c) A fourth-order $\Sigma\Delta$ modulator with multi-bit quantization, described in Chapter 7, for an ADSL (*Asymmetrical digital subscriber loop*) system that required 12-bit effective resolution at 2.2MS/s. The use of a 2-1-1 cascade and multi-bit quantization allows reduction of the oversampling ratio to only 16 (35.2-MHz clock rate). The prototype, integrated in a 0.7µm CMOS technology features 13bit at 2.2MS/s, while dissipating 55mW at 5-V supply.

As shown in Chapter 2, the use of optimization techniques in all the design levels has enabled the performance of these three modulators competitive with state-of-the-art ICs.

REFERENCES

[ADev86]   Analog Devices Inc.: *Analog-Digital Conversion Handbook* (3rd. edition). Englewood Cliffs, Prentice-Hall 1986.

[Bahe92]   H. Baher and E. Afifi: "Novel Fourth-Order Sigma-Delta Convertor". *Electronics Letters*, Vol. 28, pp. 1437-1438, July 1992.

[Bair96]   R. T. Baird and T. S. Fiez: "A Low Oversampling Ratio 14-b 500-kHz DS ADC with a Self-Calibrated Multibit DAC", *IEEE Journal of Solid-State Circuits*, Vol. 31, pp. 312-320, March 1996.

[Bose88]   B. E. Boser and B. A. Wooley: "The Design of Sigma-Delta Modulation Analog-to-Digital Converters", *IEEE Journal of Solid-State Circuits*, Vol. 23. pp. 1298-1308, December 1988.

[Bran91a] B. Brandt, D. W. Wingard and B. A. Wooley: "Second-Order Sigma-Delta Modulation for Digital-Audio Signal Acquisition", *IEEE Journal of Solid-State Circuits*, Vol. 23. pp. 618-627, April 1991.

[Bran91b] B. Brandt and B. A. Wooley: "A 50-MHz Multibit $\Sigma\Delta$ Modulator for 12-b 2-MHz A/D Conversion", *IEEE Journal of Solid-State Circuits*, Vol. 26, pp. 1746-1756, December 1991.

[Cand85]   J. C. Candy: "A Use of Double Integration in Sigma-Delta Modulation". *IEEE Transactions on Communications*, Vol. COM-33, pp. 249-258, March 1985.

[Cand92]   J. C. Candy and G. C. Temes, (Editors): "*Oversampling Delta-Sigma Converters*", IEEE Press, 1992.

[Carl89]   L. R. Carley: "A Noise-Shaping Coder Topolgy for 15+ Bit Converters", *IEEE Journal of Solid-State Circuits*, Vol. 24, pp. 267-273, April 1989.

[Chao90]   K. C. H. Chao et al.: "A Higher Order Topology for Interpolative Modulators for Oversampling A/D Converters", *IEEE Transactions on Circuits and Systems*, Vol. 37, pp. 309-318, March 1990.

[Espe96]   S. Espejo et al.: "A 0.8µm CMOS Programmable Analog-Array-

Processing Vision Chip with Local Logic and Image Memory", *in Proc. of European Solid-State Circuits Conference*, pp. 280-283, 1996.

[Ferg91] P. Ferguson, Jr. et al.: "An 18b 20KHz Dual $\Sigma\Delta$ A/D Converter", *in Proc. of IEEE International Solid-State Circuits Conference*, pp. 68-69, February 1991.

[Good96] F. Goodenough: "Analog Technologies of all Varieties Dominate ISSCC", *Electronic Design*, Vol. 44, pp. 96-111, February 1996.

[Gray87] P. R. Gray: "Analog ICs in the Submicron Era: Trends and Perspectives", *in Proc. of IEEE Electron Devices Meeting*, pp. 5-9, 1987.

[Haus86] M. W. Hauser and R. W. Brodersen: "Circuit and Technology Considerations for MOS Delta-Sigma A/D Converters", *in Proc. of IEEE International Symposium on Circuits and Systems,* May 1986.

[Haya86] T. Hayashi et al.: "A Multi-Stage Delta-Sigma Modulator without Double Integration Loop", *in Proc. of IEEE International Solid-State Circuits Conference*, pp. 182-183, February 1986.

[Inos62] H. Inose, Y. Yasuda and J. Murakami: "A Telemetering System by Code Modulation-Δ-Σ Modulation", *IRE Transactions on Space Electronics and Telemetry*, Vol. 8, pp. 204-209. September, 1962.

[Lee87] W. L. Lee and C. G. Sodini: "A Topology for Higher Order Interpolative Coders", *in Proc. of IEEE International Symposium on Circuits and Systems*, pp. 459-462, 1987.

[Leme93] C. A. Leme: *Oversampled Interfaces for IC Sensors*, ETH Press, Zurich, 1993.

[Leop91] H. Leopold, G. Winkler, P. O'Leary, K. Ilzer and J. Jernej: "A Monolitic 20 bit Analog-to-Digital Converter", *IEEE Journal of Solid-State Circuits*, Vol. 26, July 1991.

[Leun92] B. H. Leung and S. Sutarja: "Multi-bit Σ-Δ A/D Converter Incorporating A Novel Class of Dynamic Element Matching Techniques", *IEEE Transactions on Circuit and Systems-II*, Vol. 39, pp. 35-51, January 1992.

[Nish93] K.A. Nishimura: *"Optimum Partitioning of Analog and Digital Circuitry in Mixed-Signal Circuits for Signal Processing"*. PhD dissertation, Univ. of California - Berkeley, 1993.

[Nors96] S. R. Norsworthy, R. Schereier and G. C. Temes, (Editors): *"Delta-Sigma Data Converters: Theory, Design and Simulation"*, IEEE Press, 1996.

[Nys96] O. Nys and R. Henderson: "A Monolithic 19bit 800Hz Low-Power Multi-bit Sigma Delta CMOS ADC using Data Weighted Averaging", *in Proc. of European Solid-State Circuits Conference*, pp. 252-255, 1996.

[Op'T91] F. Op'T Eynde, G. Yin and W. Sansen: "A CMOS Fourth-Order 14b 500KSample/s Sigma-Delta ADC Converter", *in Proc. of IEEE International Solid-State Circuits Conference*, pp. 62-63, 1991.

[Op'T93]   F. Op'T Eynde and W. Sansen: *"Analog Interfaces for Digital Signal Processing Systems"*, Kluwer 1993.

[Plas94]   R. van de Plassche: *"Integrated Analog-to-Digital and Digital-to-Analog Converters"*, Kluwer 1994.

[Raza95]   B. Razavi: *"Principles of Data Conversion System Design"*, IEEE Press, 1995.

[Sans96]   W. Sansen: "Challenges in Analog IC Design in Submicron CMOS Technologies", *in Proc. of IEEE-CAS Workshop on Analog and Mixed IC Design*, pp. 72-78, September 1996.

[Sign90]   B. P. Del Signore D. A. Kerth, N. S. Sooch and E. J. Swanson: "A Monolithic 20-b Delta-Sigma A/D Converter", *IEEE Journal of Solid-State Circuits*, Vol. 25, pp. 1311-1317, December 1990.

[Swan95]   E.J. Swanson: "Analog VLSI Data Converters - The First 10 Years", *in Proc. of the European Solid-State Circuits Conference*, pp. 25-29, September 1995.

[Tsiv96]   Y. Tsividis: *"Mixed Analog-Digital VLSI Devices and Technology"*, MacGraw-Hill, 1996.

[Toum92]   L. Le Toumelin et al.: "A 5-V CMOS Line Controller with 16-b Audio Converters", *IEEE Journal of Solid-State Circuits*, Vol. 27, pp. 332-341, March 1992.

[Verg95]   N.K. Verghese, T.J. Schmerbeck and D.J. Allstot: *"Simulation Techniques for Mixed-Signal Coupling in Integrated Circuits"*, Kluwer 1995.

[Vitt85]   E. A. Vittoz: "The Design of High-Performance Analog Circuits on Digital CMOS Chips", *IEEE Journal of Solid-State Circuits*, Vol. 20, pp. 657-665, June 1985.

[Will91]   L. A. Williams, III and B. A. Wooley: "Third-Order Cascaded $\Sigma\Delta$ Modulators", *IEEE Transactions on Circuits and Systems*, Vol. 38, pp. 489-498, May 1991.

[Yin93]   G. M. Yin, F. Stubbe and W. Sansen: "A 16-bit 320kHz CMOS A/D Converter using 2-Stage 3rd-Order $\Sigma\Delta$ Noise-Shaping", *IEEE Journal of Solid-State Circuits*, Vol. 28. pp. 640-647, June 1993.

[Yin94]   G. Yin and W. Sansen: "A High-Frequency and High-Resolution Fourth-Order $\Sigma\Delta$ A/D Converter in Bi-CMOS Technology", *IEEE Journal of Solid-State Circuits*, Vol. 29, pp. 857-865, August 1994.

[Yuka87]   A. Yukawa: "Constraints Analysis for Oversampling A-to-D Converter Structures on VLSI Implementation", *in Proc. of IEEE International Symposium on Circuits and Systems*, pp. 467-472, 1987.

[ZCha95]   Z-Y Chang, D. Macq, D. Haspeslagh, P. Spruyt and B. Goffart: "A CMOS Analog Front-End Circuit for an FDM-Based ADSL System", *IEEE Journal of Solid-State Circuits*, Vol. 30, pp. 1449-1456, April 1995.

Chapter **2**

Oversampling Sigma-Delta A/D converters
Basic concepts and state of the art

2.1   INTRODUCTION

Oversampling converters, (or those that use a sampling frequency much larger than the Nyquist frequency of the signal being converted, [Stee75]) have become very popular during the last decade. Such success is due to the fact that they can solve some of the problems encountered in other architectures for digital CMOS implementations, mainly the need for high-selectivity analog filters, and large sensitivity to the circuitry imperfections and noisy environs.

In fact, oversampling converters relax the requirements of the analog circuitry at the expense of faster, more complex digital circuitry [Cand92]. On the one hand, the use of an unusually high sampling frequency relaxes the specifications for the antialiasing filter, so that it can be implemented with a passive first-order filter. Critical filtering is made in the digital plane where it reacts more robustly against circuit imperfections. On the other hand, by combining oversampling and Sigma-Delta (ΣΔ) modulation, [Inos62] A/D converters are generated with large resolution, robust operation and relative insensitivity to non-ideal effects.

These conveniences explain that oversampling converters, and especially those that incorporate ΣΔ modulation, are suitable for implementing A/D high-performance interfaces [Leme93] in sub-micron, low-voltage technologies, where, although it is not easy to get precise analog performance, it is clearly possible to densely implement fast digital circuits.

However, the large differences encountered, even at a conceptual level, between traditional converters and those based on ΣΔ modulation, largely make difficult the re-use of the design methodologies developed for the former. It is necessary, therefore, to develop new design methodologies that allow us to obtain optimized $\Sigma\Delta$ converters, exploiting the potential of this technique. This task is based on a deep knowledge of the modulator operation, and on how this is affected by the non-idealities of the circuitry.

This chapter is dedicated to procuring a global vision of the oversampling A/D converters using $\Sigma\Delta$ modulation. In Section 2.2 the basic concepts of oversampling, $\Sigma\Delta$ modulation and the figures of merit that characterize the performance of the $\Sigma\Delta$ A/D converters are reviewed. Section 2.3 summarizes the characteristics of several $\Sigma\Delta$ modulator architectures. Finally, Section 2.4 is dedicated to a revision of state-of-the-art A/D converters.

2.2   BASIC CONCEPTS

Fig. 2.1 shows the block diagram of an oversampling $\Sigma\Delta$ A/D converter that includes the following elements:
a) *Anti-aliasing filter.* It eliminates from the input signal the spectral components above half sampling frequency. Oversampling relaxes the requirements of this filter, so that a simple passive first-order filter suffices.
b) *Modulator.* In this block the signal is sampled and quantized. Additionally, it is possible to filter the error inherent in the quantization, by shaping its power spectral density, in such a way that most of its power lies out of the signal band, where the error is eliminated by digital filtering. This fact has resulted in the qualifier noise-shaping, that is also used to name the $\Sigma\Delta$ modulators. The modulator output is coded into a reduced number of bits (usually only one) at the sampling rate.
c) *Decimator.* In this purely digital block, after filtering all the components out of the signal band, which includes a big part of the quantization error power, data are decimated to reduce the sampling frequency up to the Nyquist frequency. The result is the signal coded in a large number of bits at the Nyquist rate.

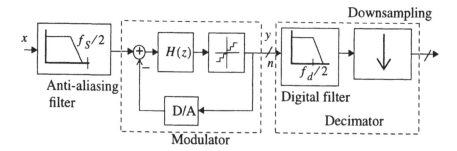

Figure 2.1: Block diagram of a $\Sigma\Delta$ A/D converter

Fig. 2.2 illustrates the processing of the signal performed by the blocks of the converter. For a better understanding of this, sampling and quantization operations have been separated in the modulator block. In the same way, filtering and decimation have been isolated in the digital part. Note that the latter does not imply a loss of information because only the redundant part of that information due to the oversampling is removed.

Among the converter blocks, the modulator is the hardest to design. On the one hand, oversampling simplifies the anti-aliasing filter up to a simple RC low-pass filter. On the other, the decimator [Nors97a] is a pure digital block whose design, like that of other DSP blocks, is highly structured and can be made with the help of CAD tools readily available. In contrast, the

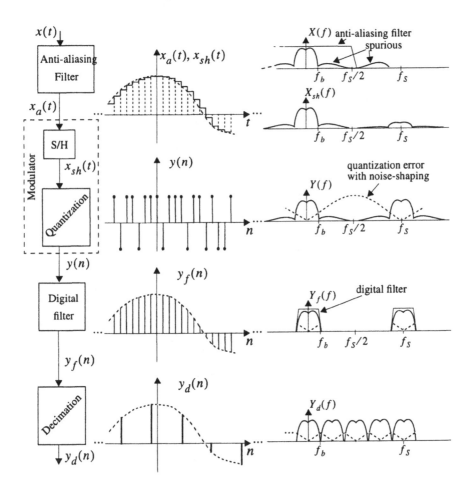

Figure 2.2: Illustrating the signal processing in a ΣΔ A/D converter

modulator, situated between the analog and digital plane, encloses many error mechanisms degrading the performance of the converter: first, quantization errors, inherent in that operation; second, a large set of circuit non-idealities that, up to a point, affect the behavior of the modulator building blocks. The impact of the latter, as well as the crosstalk from neighboring digital blocks, has to be taken into account for the design of high-performance converters.

Because of that, this work focuses on the modulator block, with the following objectives: (a) Analysis and modeling of the ΣΔ modulator behavior in presence of circuit imperfections; (b) Development of CAD tools that, incorporating the models derived, support the design tasks of such modulators; and (c) IC implementation of ΣΔ modulators in CMOS technologies.

2.2.1 Oversampling and quantization noise

Quantization in amplitude is a fundamental process in A/D conversion. Fig. 2.3 shows the transfer characteristic of an ideal quantizer. This can be represented mathematically by a non-linear function as follows,

$$y = g_q i + e \tag{2.1}$$

where g_q denotes the slope of the line intersecting the code steps or *quantizer gain*; and e stands for the quantization error. This error, inherent in that operation, is a non-linear function of the input as shown in Fig. 2.3(a). Note that, if the quantizer input is confined to the interval $[i_{min}, i_{max}]$, the quantization error is bounded by the interval $[-\Delta/2, \Delta/2]$, being Δ the separation between consecutive levels in the quantizer. For inputs out of that interval, the absolute value of the quantization error grows monotonously. This situation is known as overloading of the quantizer.

The quantization error is unambiguously determined by the input level. However, if the input varies randomly from sample to sample in the interval $[i_{min}, i_{max}]$, and the number of levels of the quantizer is large, it can be shown [Benn48] that the quantization error distributes uniformly in the range $[-\Delta/2, \Delta/2]$, with which it has a constant power spectral density, like that of white noise. Because of that, the quantization error is usually called simply *quantization noise*. As the total quantization noise power, $\sigma^2(e)$, is uniformly distributed in the range $[-f_S/2, f_S/2]$, its power spectral density equals

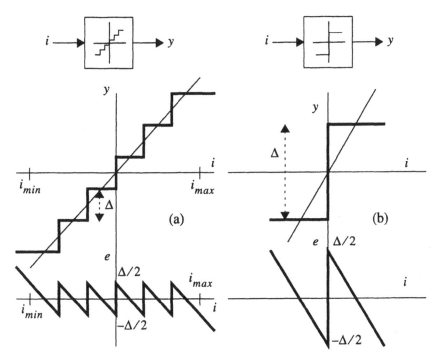

Figure 2.3: Transfer curves and quantization error of (a) a multi-bit quan
tizer; (b) a single-bit quantizer (comparator).

$$S_E(f) = \frac{\sigma^2(e)}{f_S} = \frac{1}{f_S}\left[\frac{1}{\Delta}\int_{-\Delta/2}^{\Delta/2} e^2 de\right] = \frac{\Delta^2}{12 f_S} \tag{2.2}$$

If the sampling frequency equals the Nyquist frequency, the power of the quantization error in the signal band is $\Delta^2/12$; that is, the whole noise power. But on the other hand, if the sampling frequency is larger than the Nyquist one, a lower portion of the quantization noise power remains in the signal band; this *in-band power* can be calculated as

$$P_Q = \int_{-f_d/2}^{f_d/2} S_E(f)df = \frac{\Delta^2 f_d}{12 f_S} = \frac{\Delta^2}{12M} \tag{2.3}$$

where f_d is the Nyquist frequency (twice the signal band). The in-band power of the quantization noise becomes inversely proportional to the ratio between the sampling frequency and the Nyquist frequency. This ratio, usu-ally denoted by M, is called *oversampling ratio*. According to (2.3), an increment in M implies a reduction of 3dB/octave in the in-band quantization noise power.

For single-bit quantization, Fig. 2.3(b), the above approximations do not rigorously hold; nevertheless, expertise shows that, even with these quantizers (very interesting in practice due to their simplicity) the models based on the assumption of white, additive quantization noise are approximately valid. In that case, the results shown in this section for a multi-bit quantizer are also applicable to a single-bit quantizer or comparator [Cand92].

2.2.2 $\Sigma\Delta$ Modulator

A quantizer, preceded by a sampled-and-hold block where the signal is oversampled, is the basis of the simplest oversampling converter. Let us consider a block that realizes the quantization much more efficiently: the $\Sigma\Delta$ modulator.

Fig. 2.4 shows the basic scheme of a $\Sigma\Delta$ modulator. Note that its output, y, is subtracted from its input signal, x, which has been sampled at a rate much larger than the Nyquist rate. The result, after passing through a discrete-time filter, $H(z)$, serves as an input to the quantizer itself, which usually has a reduced number of levels. If the gain of the filter is high in the interval of frequency of interest, and low out of it, the quantization error (difference between the output of the filter and that of the quantizer) is attenuated in said band due to the feedback loop.

In order to calculate the quantization error in $\Sigma\Delta$ modulators, the following considerations should be taken into account:

a) In $\Sigma\Delta$ modulators, as with an isolated quantizer, the quantization error is fully determined by the input signal.

b) The number of levels of the quantizer is usually reduced. In fact, most popular architectures incorporate a simple comparator to perform the

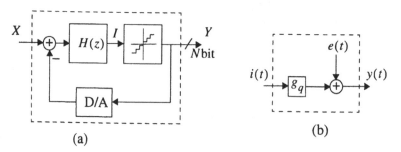

(a)

(b)

Figure 2.4: (a) Basic structure of the $\Sigma\Delta$ modulator. (b) Quantizer model.

internal quantization.

c) For static inputs, the input of the quantizer varies from sample to sample in multiples or sub-multiples of the separation between consecutive levels, Δ.

Under these circumstances, considering quantization error as non-correlated, white noise (and hence uniformly distributed in the range $[-\Delta/2, \Delta/2]$) is clearly questionable. However, when the modulator input is time-variant, previous approximations easily lead to results whose validity has been confirmed through complex calculations, including correlation between modulator input and output [Gray90].

Thus, we will assume for the quantizer the model of the Fig. 2.4(b), where $e(t)$ is uniformly distributed in the range $[-\Delta/2, \Delta/2]$, so that its power spectral density is constant and given by (2.2). Note that the gain of the quantizer in Fig. 2.4(b) only has sense when the number of levels is larger than two; otherwise, as with a comparator, the value of the gain is arbitrary because the output is just a function of the sign of the input. Once again, the linearization of the comparator transfer curve is a gross approximation that, nevertheless, leads to acceptable and mathematically simple results.

Thus, the modulator in Fig. 2.4(a) can be viewed as a two-input, $x(t)$ and $e(t)$, one-output, $y(t)$, system, which in the Z-domain can be represented by

$$Y(z) = STF(z)X(z) + NTF(z)E(z) \tag{2.4}$$

where $X(z)$ and $E(z)$ are the Z-transform of the input signal and quantization noise, respectively; and $STF(z)$ and $NTF(z)$ are the respective transfer functions of the input and quantization noise. The exact form of both functions will depend on the architecture of the modulator. Some cases will be studied here, but in view of (2.4), we can impose the following conditions to get operative modulators:

$$\begin{array}{ll} |STF(z)| = cte & \\ & \text{for} \quad z \to 1 \\ NTF(z) \to 0 & \end{array} \tag{2.5}$$

that is, quantization noise should be attenuated in the low-frequency region with no distortion of the signal.

Analyzing the scheme of Fig. 2.4, the Z-domain output yields

$$Y(z) = \frac{H(z)}{1 + H(z)}X(z) + \frac{1}{1 + H(z)}E(z) \tag{2.6}$$

Comparing this expression with that in (2.5), results in the following condition on the transfer function of the discrete-time filter:

$$H(z) \to \infty \quad \text{for} \quad z \to 1 \tag{2.7}$$

The simplest block that implements such a transfer function is an integrator, with which,

$$H(z) = \frac{z^{-1}}{1 - z^{-1}} \tag{2.8}$$

thus, substituting the filter $H(z)$ in Fig. 2.4 by a discrete-time integrator, Fig. 2.5, yields the following for the modulator output:

$$Y(z) = z^{-1}X(z) + (1 - z^{-1})E(z) \tag{2.9}$$

that is, a digital version of the delayed input, plus the quantization noise multiplied by a *shaping function*. In the time domain, the quantization error suffers a differentiation in such a way that each sample is subtracted from the previous one. Intuitively it is clear that the quantization error power decreases in the low-frequency range, where the differences from sample to sample are smaller. Note that the transfer function of the quantization noise, $NTF(z) = (1 - z^{-1})$, is of first order; so, the modulator of Fig. 2.5 is called a *first-order* $\Sigma\Delta$ modulator[1]. The only difference between the modulator of Fig. 2.5 and its conceptual representation of Fig. 2.4(a) is the inclusion of two gain factors g_1 and g_1' for the input and the feedback signal, respectively, commonly called *weights* or *gains* of the integrator. If $g_1 = g_1'$ expression (2.9) is still valid so that, in the Z-domain, the power spectral density of the shaped quantization noise is

$$S_Q(f) = S_E(f) \left| 1 - \exp\left(-j2\pi \frac{f}{f_S}\right) \right|^2 = S_E(f) \cdot 4\sin^2\left(\pi \frac{f}{f_S}\right) \tag{2.10}$$

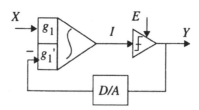

Figure 2.5: First-order $\Sigma\Delta$ modulator

---

1. An isolated quantizer can be considered as a zero-order modulator.

and its in-band power is calculated as follows:

$$P_Q = \int_{-f_d/2}^{f_d/2} S_Q(f)df \cong \frac{\Delta^2}{12}\frac{\pi^2}{3M^3} \qquad M = \frac{f_s}{f_d} \gg 1 \qquad (2.11)$$

In this case, an increase of the oversampling ratio leads to a decrease of 9dB/octave of the in-band noise power, 6dB/octave more than for a simple quantizer.

2.2.3   Signal-to-noise ratio, dynamic range and equivalent resolution

It is convenient to define at this point the figures of merit commonly used to characterize the oversampling converters.

2.2.3.1 Signal-to-noise ratio (*SNR* or *S/N*)

This is the ratio between the output power at the frequency of a sinusoidal input and the in-band noise power. It is usually given in decibels. Ideally, that is, with the quantization noise only, the *SNR*, which depends on the input amplitude A, results in:

$$SNR(dB) = 10\log_{10}\left(\frac{A^2/2}{P_Q}\right) \qquad (2.12)$$

Note that the *SNR* monotonously increases with the input level. However, beyond a certain input level, the quantizer input lies out of the interval $[i_{min}, i_{max}]$, defined in Section 2.2.1, which produces the overloading of the latter and consequently a sharp drop is observed in the *SNR* curve.

As will be shown in Chapter 3, besides quantization noise, there are other contributions to the in-band noise power due to non-idealities of the circuitry. To take into account all these errors, the signal-to-(noise distortion) ratio (*TSNR* or *S/(N+D)*) is normally used.

2.2.3.2 Dynamic range (*DR*)

The dynamic range is defined as the ratio between the output power at the frequency of a sinusoidal input with full-scale range amplitude and the output power when the input is a sinusoide of the same frequency, but of a small amplitude, so that it cannot be distinguished from noise; that is, with $SNR = 0$dB. It is usually given in dB.

Ideally, the full-scale range of the modulator input is approximately given by that of the quantizer. In a single-bit quantization case, this range equals $\pm\Delta/2$, so the full-scale input amplitude is $\Delta/2$. On the other hand, according to (2.12) the output power for such an input as $SNR = 0$dB is P_Q; so,

$$DR(\text{dB}) = 10\log_{10}\left[\frac{(\Delta/2)^2}{2P_Q}\right] \tag{2.13}$$

For the first-order modulator (Fig. 2.5), we obtain

$$DR(\text{dB}) = 10\log_{10}\left(\frac{9}{2}\frac{M^3}{\pi^2}\right) \tag{2.14}$$

2.2.3.3 Effective resolution (B)

The dynamic range of an ideal B-bit A/D converter is given by [Bose88b]

$$DR = 3 \cdot 2^{(2B-1)} \tag{2.15}$$

Manipulating this expression yields the effective number of bits or effective resolution of the $\Sigma\Delta$ modulator as a function of its DR (dB),

$$B(\text{bit}) = \frac{DR(\text{dB}) - 1.76}{6.02} \tag{2.16}$$

Note that a 3-dB increase in DR implies a 0.5-bit increase in effective resolution, while a 9-dB increase leads to a 1.5-bit extra resolution. Therefore, increasing the oversampling ratio leads to an increase in effective resolution of 0.5bit/octave for a single quantizer. This increment rises to 1.5bit/octave for a first-order $\Sigma\Delta$ modulator. Another interesting formula relates the achievable effective resolution to the oversampling ratio and internal quantization resolution (b) as follows:

$$B(\text{bit}) = \frac{1}{2}\log_2\left[\frac{(2^b-1)^2(2L+1)M^{2L+1}}{\pi^{2L}}\right] \tag{2.17}$$

where L denotes the modulator order.

2.2.4 Pattern-noise generation

Expressions derived for the in-band quantization noise have, as a premise, that the samples of the quantization error and those of the signals are non-correlated, which is approximately valid when the input varies over time [Benn48]. Nevertheless, as shown in Appendix A, for static inputs, the output of a first-order ΣΔ modulator tries to equal on average the input level with repetitive patterns [Cand81]. Such periodicity makes the quantization noise not white but intensely colored. The in-band power of the quantization error (which cannot be called noise in this case) is then rather larger than that predicted in (2.11). This difference is apparent when the frequency of the patterns lies in the signal band, which mainly occurs when the input level is close either to zero or to the reference voltage. For DC inputs in $[-\Delta/2, \Delta/2]$, the in-band quantization error power sharply changes with the input, as shown in Fig. 2.6 [Cand92]. A possible solution to this problem is to include, usually at the quantizer input, some non-periodic signals, as pseudo-random noise, for example. With this technique (called *dittering* [Nors97b]) it is possible to partially decorrelate the quantization error and the input, at the price of larger complexity of the design. Besides, the dittering signal itself is a noise source at the modulator output. Another possibility to decorrelate signal and quantization error is to use some chaotic behavior in the modulator [Schr93]. Such chaos is obtained in practice by moving some of the zeros of $NTF(z)$ out of the unity circle. As a drawback, a tendency to instability has been noticed when the number of quantization levels is reduced [Nors97b].

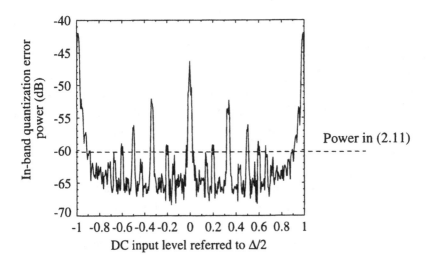

Figure 2.6: Pattern noise for a first-order ΣΔ modulator with $M = 64$

2.3   SIGMA-DELTA MODULATOR ARCHITECTURES

Since the middle-eighties a large amount of $\Sigma\Delta$ modulator architectures based on that of Fig. 2.4(a) have been reported. In all cases a reduction in the in-band power of the quantization noise is pursued by following two different, non-exclusive strategies:

a) Increasing the order of the filter $H(z)$, which leads to an increase of the order of $NTF(z)$, resulting in a more effective cancellation of the quantization noise. In such architectures the extra resolution when the oversampling ratio is increased is well above 1.5bit/octave.

b) Augmenting the resolution of the internal quantizer; with that, Δ and hence the power spectral density of the quantization noise is reduced. In fact, ideally on average each extra bit in the internal quantization results in an extra bit of the effective resolution of the modulator.

By adopting one of these strategies or by means of a combination of both, $\Sigma\Delta$ modulator architectures have been conceived that require a lower oversampling ratio to get a given resolution, which provides lower power-speed ratio. An exhaustive study of all possible architectures goes beyond the intentions of this book. Instead of that, a summary of the main characteristics of the most used architectures is given next. Special emphasis is placed on those that will be used for practical implementations.

2.3.1   Second-order $\Sigma\Delta$ modulator

Including one more integrator in the modulator loop increases the noise transfer function order to two. The resulting architecture is the single-loop second-order $\Sigma\Delta$ modulator [Cand85] of Fig. 2.7.

The time-domain analysis of this modulator is rather complex due to the double integration[†1]. However, such analysis is simplified in the Z-domain. With $g_2' = 2g_1g_2$, which guarantees the stability of the loop [Cand85], and $g_1' = g_1$ the following is obtained for the modulator output:

$$Y(z) = z^{-2}X(z) + (1 - z^{-1})^2 E(z) \tag{2.18}$$

where it has been assumed that both integrators have the transfer function in (2.8) and that the quantization error is additive. Comparing this expression to (2.9) note that the double integration has augmented by one the order of $NTF(z)$. Consequently, the power spectral density of the quantization noise

---

1. An analysis of the time-domain response of 1st- and 2nd-order modulators is given in Appendix A.

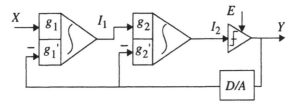

Figure 2.7: Single-loop second-order ΣΔ modulator

yields:

$$S_Q(f) = S_E(f)\left|1 - \exp\left(-j2\pi\frac{f}{f_S}\right)\right|^4 = S_E(f) \cdot 16\sin^4\left(\pi\frac{f}{f_S}\right) \tag{2.19}$$

Fig. 2.8 points out the advantages of the second-order modulator with respect to the first-order one, by comparing both transfer functions for the quantization noise in the frequency domain. Note that for the second-order modulator, the spectral density significantly diminishes in the low-frequency region, at the price of an increase in the high-frequency region. Displacing this power leads to a decrease of the in-band quantization noise, calculated as follows:

$$P_Q = \int_{-f_d/2}^{f_d/2} S_Q(f)df \cong \frac{\Delta^2}{12}\frac{\pi^4}{5M^5} \qquad M = \frac{f_S}{f_d} \gg 1 \tag{2.20}$$

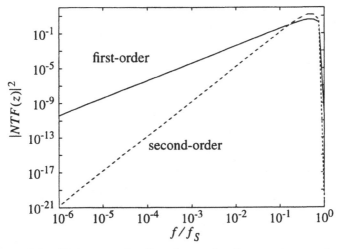

Figure 2.8: Noise transfer functions of a first- and second-order
ΣΔ modulator as a function of the frequency

Thus, an increase of the oversampling ratio causes a 15dB/octave decrease in the in-band noise or a 2.5bit/octave increase in effective resolution, instead of the 1.5bit/octave obtained with a first-order $\Sigma\Delta$ modulator (2.11). For instance, ideally, the dynamic range provided by a second-order modulator is 34dB higher than that provided by a first-order one for $M = 128$; this difference increases to 40dB if $M = 256$, which corresponds to 5.7 and 6.7-bit extra resolution, respectively.

In addition to the advantage regarding the achievable dynamic range, the second-order modulator shows smaller noise patterns in the presence of static signals [Cand85]. This reduction, shown in Fig. 2.9 for $M = 64$, is obtained thanks to the use of double integration which partially decorrelates the input signal and the quantization error (see Appendix A).

In fact, the presence of idle tones in the signal band decreases as the modulator order increases. In practice, this fact, together with the presence of electronic noise in circuit devices, which acts as a dittering signal, enables us to largely obviate the problem of the quantization noise coloring.

2.3.2 Single-loop high-order modulators

The above expression for a 1st- and 2nd-order $\Sigma\Delta$ modulator can be extended to a modulator of order L, whose more direct implementation consists of including L integrators before the quantizer [Ritc77]. The resulting architecture is shown in Fig. 2.10. The Z-domain modulator output is a generalization of the expressions (2.9) and (2.18):

$$Y(z) = z^{-L}X(z) + (1 - z^{-1})^L E(z) \tag{2.21}$$

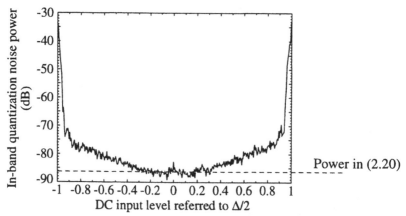

Figure 2.9: Pattern noise of a 2nd-order $\Sigma\Delta$ modulator with $M = 64$

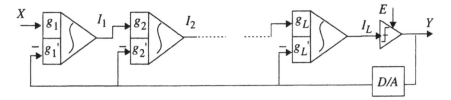

Figure 2.10: Lth-order single-loop ΣΔ modulator

with which the spectral density of the quantization noise is given by:

$$S_Q(f) = S_E(f)\left|1 - \exp\left(-j2\pi\frac{f}{f_S}\right)\right|^{2L} = S_E(f) \cdot 2^{2L}\sin^{2L}\left(\pi\frac{f}{f_S}\right) \qquad (2.22)$$

Integrating (2.22) in the signal band $[-f_d/2, f_d/2]$ results in the following in-band noise power:

$$P_Q = \int_{-f_d/2}^{f_d/2} S_Q(f)df \cong \frac{\Delta^2}{12}\frac{\pi^{2L}}{(2L+1)M^{(2L+1)}} \qquad M = \frac{f_S}{f_d} \gg 1 \qquad (2.23)$$

In general, doubling the oversampling ratio results in a reduction of $3(2L+1)$dB in the in-band quantization noise power, which equals to $L + 1/2$ bit/octave extra effective resolution.

2.3.2.1 Stability considerations

A drawback of the ΣΔ modulators with $L > 2$ is their tendency to instability [Op'T93]. A modulator is considered stable if, for bounded inputs and whatever integrator initial conditions, the internal state variables (integrator outputs) remain also bounded over time. It can be shown that a first-order modulator is intrinsically stable for whatever input in the range $(-\Delta/2, \Delta/2)$. In the same way, the stability of the second-order modulator of Fig. 2.7, for whatever input in the range $(-0.9\Delta/2, 0.9\Delta/2)$, is guaranteed if $g_2'/(g_1g_2) > 1.25$ [Cand85].

However, when the order is larger than two, it is not possible to mathematically determine a stability condition[1]. Using behavioral simulation, it has been shown [Op'T93] that, with proper selection of the scaling coefficients,

---

1. In [Goods95] the study is extended to the static-input third-order modulator case by using the concept of invariant sets. However, the application of this method to higher-order modulators or with dynamic inputs is extremely complex.

the modulator with an order larger than two is conditionally stable; that is, stable operation is only obtained for inputs confined to a given interval, which depends on the modulator architecture and on the integrator initial conditions. For inputs or initial conditions out of their respective ranges, these modulators are unstable. Additionally, as opposed to the low-order modulator, the oscillations due to an excessive transient input, like that eventually produced after circuit power-on, do not disappear even though the signal becomes bounded in the stable input range after the transient.

In practice, stable high-order modulators can be obtained by using several techniques:

a) Proper selection of the scaling coefficients (integrator weights), in order to reduce the out-of-the-band gain for the quantization noise to a level that ensures the stability of the loop.

b) Taking advantage of the bounded nature of the signals at the integrator outputs due to the requirements of the circuitry, or including limiters after the last integrators in the chain.

c) Global resetting of the integrators when an unstable operation is detected. The detection of instability can be done at the integrator level, by placing comparators to determine whether an internal state variable has surpassed a certain limit, or by monitoring the length of the series of consecutive pulses at the modulator output.

In [Adam97a] a summary of these techniques is made in the form of a set of recipes for the design of high-order stable modulators. According to the author´s conclusions, the resulting modulators have an SNR-peak rather smaller than that calculated with (2.17). For instance, a fifth-order modulator with M = 64 stabilized with the above techniques loses 60dB in respect to the ideal case. Moussavi and Leung [Mous94] have recently proposed the selective re-setting of integrators with excessive output, using comparators and local feed-back. The outputs of these comparators are digitally processed, together with the modulator output, in such a way that the error due to the re-setting can be compensated. However, the third-order modulator in [Mous94], using this technique has an SNR-peak of 84dB for M = 64; that is, 28dB smaller than that predicted by (2.17).

2.3.3 High-order $\Sigma\Delta$ modulator architectures

In addition to the architecture of Fig. 2.10, there are other possibilities for obtaining high-order noise-shaping functions [Ribn91]. Some of them are considered next.

2.3.3.1 Single-loop modulators

The easiest way to obtain Lth-order $NTF(z)$ is an Lth-order differentiation, (2.21). Nevertheless, the value of such a function at a frequency near to half the sampling frequency quickly increases with L, providing modulators with a clear tendency to instability. A solution to this problem was proposed by Lee and Sodini [Lee87] with an architecture that makes possible the generation of $NTF(z)$ with multiple poles and zeros along the signal-band and much smaller gain out of that band[†1]. The Lee-Sodini architecture is shown in Fig. 2.11. Its Z-domain response is given by (2.4) with, [Lee87].

$$STF(z) = \frac{\displaystyle\sum_{i=0}^{L} A_i(z-1)^{N-i}}{z\left[(z-1)^L - \displaystyle\sum_{i=1}^{L} B_i(z-1)^{L-i}\right] + \displaystyle\sum_{i=0}^{L} A_i(z-1)^{L-i}} \qquad (2.24)$$

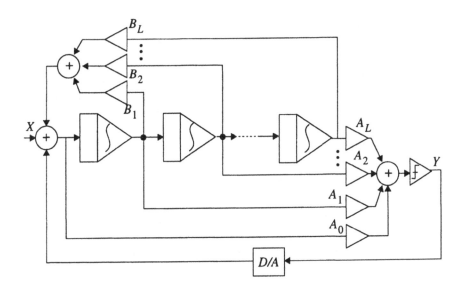

Figure 2.11: Lee-Sodini Lth-order ΣΔ modulator

---

1. A similar architecture using a multi-bit quantizer was proposed by Carley [Carl87].

$$NTF(z) = \frac{(z-1)^L - \sum_{i=1}^{L} B_i(z-1)^{L-i}}{z\left[(z-1)^L - \sum_{i=1}^{L} B_i(z-1)^{L-i}\right] + \sum_{i=0}^{L} A_i(z-1)^{L-i}} \tag{2.25}$$

If all B_i coefficients are zero, $NTF(z)$ has all zeros located at DC. The high-pass function is thus similar to that obtained with a Butherworth or Chebychev filter. Otherwise, if B_is are adjusted to place the zeros in the stop-band, the selectivity of the filter is maximized, and consequently, the in-band quantization noise power minimized, with characteristics similar to an inverse Chebychev or elliptic filter.

Apart from the Lee-Sodini modulator, also called interpolative, there exist other possibilities for obtaining high-order noise-shaping functions, with the characteristics of most used filters [Adam97b]. Most of them use feedback or feedforward transmission of the signals in the loop. A drawback of these modulators is the increased complexity of the analog circuitry. Also, the need for very small coefficients implies the use of very large capacitor values in an SC implementation, which increases the power consumption.

2.3.3.2 Cascade modulators

An alternative to the single-loop or interpolative modulators are the cascade architectures (multi-stage or MASH) [Chou89][Long88][Mats87] [Rebe90] whose generic block diagram is shown in Fig. 2.12. Their functioning is based on the cascade connection of low-order modulators (0, 1 or 2), whose stability is guaranteed by design. The quantization noise generated in one stage is then re-modulated by the next one, and later cancelled in the digital domain. As a result, ideally, one obtains the modulator input plus the quantization noise of the last stage, attenuated by a shaping function of order equal to the number of integrators in the cascade.

As a matter of example, let us consider the $\Sigma\Delta$ modulator of Fig. 2.13 [Yin93b]. This architecture is a possible implementation of a fourth-order cascade with the architecture 2-1-1; that is, a second-order modulator as the first stage followed by two first-order modulators. Its Z-domain output can be represented by

$$Y(z) = STF(z)X(z) + NTF_1(z)E_1(z) + NTF_2(z)E_2(z) + NTF_3(z)E_3(z) \tag{2.26}$$

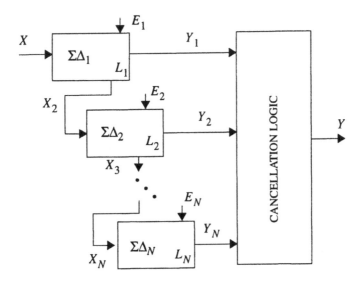

Figure 2.12: Block diagram of a cascade ΣΔ modulator

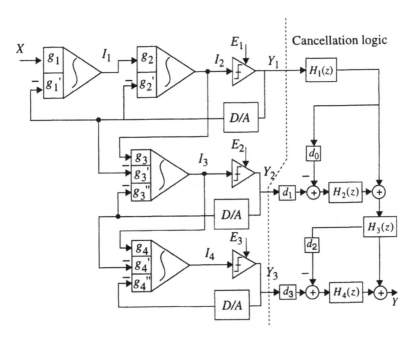

Figure 2.13: 4th-order 2-1-1 cascade ΣΔ modulator

According to the above considerations, to obtain the performance of a fourth-order modulator, the following relationships must be met:

$$STF(z) \sim z^{-4}$$
$$NTF_1(z), NTF_2(z) = 0$$
$$NTF_3(z) \sim (1 - z^{-1})^4$$

(2.27)

which translates into the relationships among coefficients and $H_i(z)$ transfer functions shown in Table 2.1.

Table 2.1: Relationships for the 2-1-1 architecture

Analog	Digital/Analog		Digital
$g_1' = g_1$	$d_0 = 1 - g_3'/(g_1g_2g_3)$		$H_1(z) = z^{-1}$
$g_2' = 2g_1'g_2$	$d_1 = g_3''/(g_1g_2g_3)$		$H_2(z) = (1 - z^{-1})^2$
$g_4' = g_3''g_4$	$d_2 = \left(1 - \dfrac{g_3'}{g_1g_2g_3}\right)\left(1 - \dfrac{g_4'}{g_3''g_4}\right) \equiv 0$		$H_3(z) = z^{-1}$
	$d_3 = g_4''/(g_1g_2g_3g_4)$		$H_4(z) = (1 - z^{-1})^3$

With which

$$Y(z) = z^{-4}X(z) + d_3(1 - z^{-1})^4 E_3(z)$$

(2.28)

Note that, in general, due to the need for scaling the signals along the cascade, the coefficient d_3 (see Table 2.1) will be larger than the unity. This means an increment of the quantization noise and hence a systematic loss of dynamic range. However, in practice, a proper choice of the analog coefficients reduces this loss to only 6dB.

Thus, with this technique it is possible to obtain high-order, stable $\Sigma\Delta$ modulators with *SNR* values close to those predicted in (2.17). As a drawback, they exhibit larger sensitivity to certain non-ideal aspects of the circuitry. The impact of such non-idealities on the performance of a cascade modulator is analyzed in Chapter 3. A comparative study of a 4th-order cascade modulator is given in Chapter 6.

2.3.4   Multi-bit quantization ΣΔ modulators

As stated in (2.17), another possibility for increasing the effective resolution of ΣΔ modulators is to increase the number of levels of the internal quantization [Paul87][Carl97]. These converters, whose generic architecture coincides with that of Fig. 2.4(a), have important advantages: the previously mentioned resolution increase does not depend on the oversampling ratio yielding a minimum of 6dB reduction in quantization noise per extra bit of the internal quantizer; the use of multi-bit quantization improves the stability of high-order loops because it is easier to anticipate the saturation of the quantizer; and finally the larger the number of levels in the quantizer, the more exact the approximate analysis based on the linearization of its transfer curve will be.

However, without proper techniques, the linearity of a multi-bit ΣΔ modulator is limited by that of the D/A converter needed in the feedback path [Adam91a][Plass79][†1]. Chapter 7 is devoted to the study of multi-bit SC ΣΔ modulators, presenting some techniques to palliate the sensitivity to the non-linearity.

To conclude, Fig. 2.14 shows a classification of the most used ΣΔ modulator architectures. Such a classification is not intended to be an exhaustive compilation of all possible topologies but to show concisely the trends previously mentioned. Also, the advantages and drawbacks of each possibility are summarized.

---

1. Note that this problem would not appear if the quantization were single bit, because, that being the case, the D/A converter would be perfectly linear per construction.

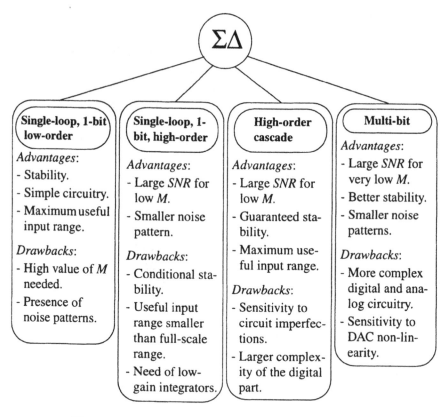

Figure 2.14: Summary of ΣΔ modulator architectures

2.4 STATE-OF-THE-ART A/D CONVERTERS

Most ΣΔ modulator architectures in the previous section have been used for the monolithic implementation of A/D converters in several technologies (CMOS, BiCMOS, Bipolar), with diverse circuit techniques: switched capacitors (SC), switched currents, (SI), continuous time, etc., and gradually decreasing the supply voltage, from 10 to 1V.

A detailed compilation of all those realizations is, due to the excessive volume of information, of little practical interest. Instead, Table 2.2 summarizes the A/D converter ICs published in the period 1991-1997. For each of them, the effective resolution, the digital output rate (*DOR*, which equals the Nyquist frequency of the signal), the power consumption, the characteristics of the fabrication process and the modulator architecture are indicated. Most ΣΔ modulators in Table 2.2 have been implemented using SC circuits.

Table 2.2: Summary of ΣΔ A/D converters published in the period 1991-1997

Ref.	Bit	DOR (kS/s)	Power (mW)	Process / Supply	Architecture
[Bert93]	21	DC	6	2μm CMOS / 5V	1st-order + PWM
[Bran91a]	16	50	13.8	1μm CMOS /5V	2nd-order
[Gril96]	15.3	6	2	0.6 μm CMOS / 1.8V	2nd-order
[Pelu96]	12	6	0.1	0.7μm CMOS / 1.5V	2nd-order
[Send97]	14.3	20	0.55	0.5μm CMOS / 1.5V	2nd-order, double sampling
[Tan95a]	11	20	2	0.8μm CMOS / 3.3V	2nd-order SI
[Tan95b]	10	15.6	0.78	0.8μm CMOS / 1.2V	2nd-order SI
[Nys96]	19	0.8	1.35	2 μm CMOS / 5V	2nd-order, 3-bit
[Saue95]	12	32	1	1.5μm CMOS / 2.3V	3rd-order
[Au97]	11.8	15.6	0.34	1.2μm CMOS / 1.95V	3rd-order, local feedback
[Kert94]	20	0.8	50	3 μm CMOS / 10V	4th-order
[Zwan97]	15.7	40	2.3	0.8μm CMOS / 3.3V	gm-C, 4th-order
[Bair96]	13.7	500	58	1.2μm CMOS / 5V	4th-order, 4-bit
[Thom94]	20	0.984	45	2 μm CMOS / 10V	5th-order (tri-level)
[Mino95]	11	200	94	0.8μm CMOS / 3V	7th-order, feedforward
[Leun97]	19.3	44	760	0.8μm CMOS / 5V	7th-order, 3-level (digitally corrected)
[Yin93a]	15.7	320	65	1.2μm CMOS / 5V	2-1 cascade
[Will94]	17	50	47	1μm CMOS /5V	2-1 cascade
[Rabi96]	15	50	5.4	1.2 μm CMOS / 1.8V	2-1 cascade
[Bran91b]	12	2100	41	1μm CMOS / 5V	3-bit, 2-1 cascade
[Yin94]	15.8	1500	180	2μm BiCMOS / 5V	2-1-1 cascade
[Fuji97]	18	48	500	0.7μm CMOS / 5V	tri-level, 2-2 cascade
[Dedi94]	14.7	200	40	1.2μm CMOS /5V	tri-level, 2-2-2 cascade
[Mats94]	9.4	384	1.56	0.5μm CMOS / 1V	RC, swing-suppression

Only those in [Tan95a][Tan95b] use SI circuits and [Zwan97] uses continuous-time gm-C integrators. The predominance of SC circuits in respect of other circuit techniques, apparent in literature, is a consequence of their larger robustness against circuitry imperfections. Modulators using current-mode basic cells, promising because of their compatibility with CMOS digital processes, have not yet reached the performance of those employing SC circuits. In addition, with the exception of the modulator in [Yin94], all fabrication processes are CMOS. This table also shows the present trend to reduced supply voltage up to 2V and below [Au97][Gril96][Mats94][Pelu96] [Saue95][Send97][Rabi96] [Tan95b].

In Fig. 2.15, the converters in Table 2.2 are placed in the resolution-speed plane, Fig. 2.15(a), and in the power consumption-speed plane, Fig. 2.15(b). For completeness, some high-frequency Nyquist A/D converters (that is, those that do not use oversampling) implemented with diverse architectures (flash, pipeline, successive approximations, etc.) are also included. The performance of these converters is shown in Table 2.3.

Table 2.3: Summary of other recently published Nyquist converters

Ref.	Bit	DOR (ks/s)	Power (mW)	Process / Supply	Architecture
[Hamm97]	9.1	200	12	1µm CMOS / 5V	Successive approximations
[Shu95]	11	10000	360	1.4µm BiCMOS / 5V	Pipelined, $\Sigma\Delta$ calibration
[Song95]	8.7	20000	50	1.2µm CMOS / 5V	Pipelined
[Wu95]	8.2	4500	128	0.8µm CMOS / 5V	Pipelined SI
[Mant96]	9.4	300	5.4	0.8µm CMOS / 2.7V	Interleaved pipelined
[Ahm96]	10	10000	250	0.8µm CMOS / 5V	Pipelined
[Lim96]	11	10000	250	0.8µm CMOS / 5V	Pipelined
[Clin96]	13	5000	166	1.2µm CMOS / 5V	Pipelined
[Yu96]	11	5000	33	1.2µm CMOS / 2.5V	Pipelined
[Kwak97]	14.1	5000	60	1.4µm CMOS / 5V	Pipelined
[Ito94]	9	20000	135	0.8µm CMOS / 3V	Subranging
[Yots95]	9	20000	20	0.5µm CMOS / 2V	Mixed-mode subranging
[Vene96]	7.1	80000	80	0.5µm CMOS / 3.3V	Folding
[Bult97]	8.7	48000	170	0.5µm CMOS / 5V	Flash + Folding
[Spal96]	5	200000	400	0.6µm CMOS / 5V	Flash

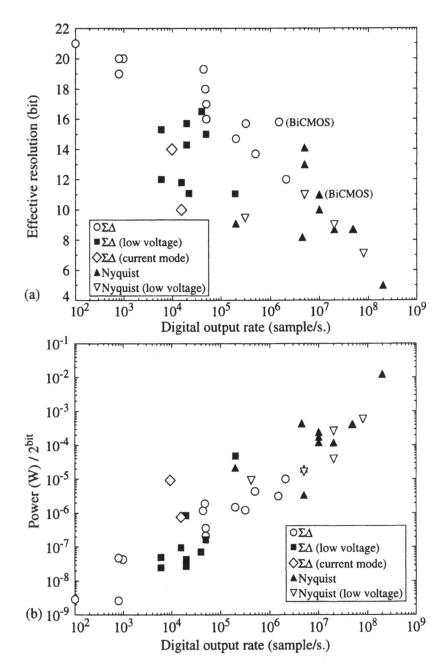

Figure 2.15: (a) Effective resolution against *DOR*. (b) Power consumption normalized to the number of bin, according to the equivalent resolution, as a function of the oversampling ratio.

Regarding Fig. 2.15(a), the extension of the application range of the over-sampling converters from DC to 2MS/s should be noted. At higher frequencies, only Nyquist converters have been reported.

Fig. 2.15(b) shows the power consumption normalized to the number of levels of a hypothetic quantizer with the same effective resolution (2^{bit}), which on average increases with the digital output rate of the modulator [Malo95][Nys93]. Note that the SI modulators dissipate more power than the SC modulators in the same frequency range. On the other hand, the power consumption is smaller in low-voltage modulators, which concentrate in the audio range: although they reach a smaller effective resolution in respect to the 5-V modulators, Fig. 2.15(a), the reduction of the supply voltage and the use of low-power specific circuitry [Rodr95] allow a significant reduction of the power consumption.

2.4.1   Comparative evaluation of reported $\Sigma\Delta$ modulators

Recently, a *Figure Of Merit* (*FOM*) has been proposed for comparative evaluation of data converter integrated circuits. The original formula of the *FOM* [Good96] can be adapted for $\Sigma\Delta$ modulators as follows,

$$FOM = \frac{Power(\text{W})}{2^{resolution(\text{bit})} \times DOR(\text{S/s})} \times 10^{12} \qquad (2.29)$$

where *resolution* denotes the effective number of bits and *DOR* is the digital output rate that coincides with twice the signal band, f_d. The result (in pico-joules) is the energy needed per conversion. The power consumption in (2.29) corresponds only to the modulator. Unfortunately, in most of the published works only the modulator is implemented, while the digital part of the $\Sigma\Delta$ converter is implemented by software, so there are hardly any data about the power consumption of the digital filter. An estimation of such consumption is complex: although it is true that the digital circuitry that must accompany a high-order or multi-bit modulator is more complex than that of a low-order, single-bit one, it is also true that the former will require a lower over-sampling ratio and hence a smaller sampling frequency to achieve a similar performance. Both effects may roughly be compensated in such a way that it is possible to comparatively evaluate rather different modulator architectures without taking into account the power consumption of the digital part[†1].

The objective of the reduction of the power consumption translates into

---

1. However, the power consumption of the digital part must be taken into account to realistically compare oversampled with non-oversampled A/D converters [Good96].

the minimization of the *FOM*. As a reference, Table 2.4 shows the sub-set of modulators in Table 2.2 with *FOM* smaller than 10pJ, ordered according to this value.

Table 2.4: Summary of ΣΔ modulators with *FOM*<10

Reference	Resolution (bits)	*DOR* (kS/s)	Power (mW)	Technology	Architecture	*FOM* (pJ)
[Send97]	14.3	20	0.55	0.5µm CMOS / 1.5V	2nd -order	1.4
[Zwan97]	16.5	40	2.3	0.8µm CMOS / 3.3V	gm-C, 4th-order	1.8
[Yin94]	15.8	1500	180	2µm BiCMOS / 5V	2-1-1 cascade	2.1
[Nys96]	19	0.8	1.35[a]	2 µm CMOS / 5V	2nd-order, 3-bit	3.2
[Rabi96]	15	50	5.4	1.2 µm CMOS / 1.8V	2-1 cascade	3.3
[Yin93a]	15.7	320	65	1.2µm CMOS / 5V	2-1 cascade	3.9
[Pelu96]	12	6	0.1	0.7µm CMOS / 1.5V	2nd-order	4.1
[Bran91a]	16	50	13.8	1µm CMOS /5V	2nd-order	4.3
[Bran91b]	12	2100	41	1µm CMOS / 5V	2-1 cascade, 3-bit	4.8
[Mats94]	9.4	384	1.56	0.5µm CMOS / 1V	RC, swing-suppression	6.0
[Will94]	17	50	47	1µm CMOS /5V	2-1 cascade	7.2
[Dedi94]	14.7	200	40	1.2µm CMOS /5V	2-2-2 cascade (3-level)	7.7
[Gril96]	15.3	6	2	0.6 µm CMOS / 1.8V	2nd-order	8.1
[Bair95]	13.7	500	58	1.2µm CMOS / 5V	4th-order, 4-bit	9.0

a. Although this ΣΔ converter includes the digital filtering, the power consumption here corresponds only to the modulator.

Note that the smallest FOM reported so far is 1.4pJ and corresponds to a second-order architecture with 1.5-V supply working in the audio range [Send97]. At higher frequencies, the best FOM is 2.1pJ [Yin94] obtained with a 2-1-1 cascade in a BiCMOS technology. The same authors reported a 2-1 cascade capable of operating at 320kS/s with a FOM of 3.9pJ [Yin93a]. The largest-*DOR* modulator in Table 2.4 (2.1MS/s) [Bran91b] uses a cascade architecture with multi-bit quantization to yield 4.8pJ.

2.4.2 Modulators in this book with respect to the state-of-the-art design

The results of this work are demonstrated through the implementation of three ΣΔ modulators in CMOS technologies. These modulators, together with the methodology that has enabled their design, have been developed inside two CEE-ESPRIT projects: #5056 (AD2000) and #8795 (AMFIS) dedicated to basic research on high-performance CMOS data converters and their design methodologies (AD2000), and to their application to diverse specific industrial products (AMFIS). In particular, two of the modulators presented here are directly related to two of the industrial applications in AMFIS.

The second-order modulator described in Chapter 5, was designed as a part of an energy metering system front end. Experimental results show that the prototype has an effective resolution of 16.4bit at $DOR = 9.6$kS/s, which is enough for the energy metering requirements. In addition, the modulator dissipates only 1.71mW operating a 5-V supply and at the nominal clock frequency (2.5MHz). This performance means a FOM of 2pJ, which places this modulator among those at the top of Table 2.4. Only two very recent low-voltage modulators [Send97][Zwan97], have lower FOM.

The fourth-order $\Sigma\Delta$ modulator with multi-bit quantization described in Chapter 7 was designed to be incorporated into an ADSL high-speed communication system over copper wire [ZCha95] which required 12-bit effective resolution at 2.2MS/s. The use of a high-order cascade architecture and dual (single-bit and 3-bit) quantization allowed reduction of the oversampling to only 16. Measurements show 13-bit effective resolution at 2.2MS/s with a power dissipation of 55mW, which translates to a FOM of 3.1pJ; that is, 1.7pJ lower than the modulator, similar in performance, in [Bran91b].

Finally, the fourth-order modulator described in Chapter 6 was designed as a demonstrator of the AD2000 project. The initial specifications (17-bit at 40kS/s) were met with a 2-2 cascade architecture, whose design, in order to evaluate the effectiveness of the methodology developed, was made in such a way that the expected resolution was exactly the required one. Experimental results are 16.7-bit effective resolution at 40kS/s dissipating 10mW from a 5-V supply. This performance represents 2.3-pJ FOM, only 0.5pJ larger than the modulator in [Zwan97], reported very recently.

These three modulators are placed in the resolution-speed plane in Fig. 2.16(a) and in the power consumption-speed plane in Fig. 2.16(b) together with those in Table 2.2. Regarding the former, it is worth mentioning the position of the cascade multi-bit modulator which is the largest-DOR $\Sigma\Delta$ modulator reported until now. On the other hand, note that the three modulators appear in the power consumption-speed plane below the best fitting straight line of the 5-V supply CMOS $\Sigma\Delta$ modulators in Table 2.2.

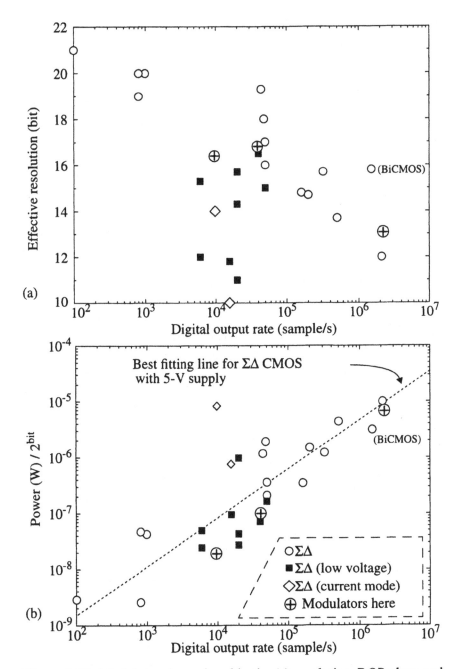

Figure 2.16: Modulators here placed in the (a) resolution-*DOR* plane and (b) power (W) / 2^{bit}- *DOR* plane.

SUMMARY

In this chapter, $\Sigma\Delta$ modulation techniques for A/D conversion and their related basic concepts have been introduced. In this context, as in the rest of this book, special attention is paid to the modulator because it is in this block that the most degrading error mechanisms are produced.

In particular, this chapter concentrated on the analysis of an error mechanism that takes place in any $\Sigma\Delta$ modulator: the quantization noise. The incidence of such an error, which can be considered as ideal since it is inherent to the quantization procedure, is attenuated in $\Sigma\Delta$ modulators through the combined use of *oversampling* and *noise-shaping*. The latter allows filtering of the quantization noise in such a way that most of its power lies out of the signal band. In fact, the larger the ratio between the sampling and the Nyquist frequency (*oversampling ratio*), or the larger the *order* of the modulator, the more efficient the filtering will be.

From this point of view, the basic architectures (first- and second-order modulators) have been briefly analyzed. Also, the main features of other more complex modulator architectures have been summarized. For all of these, the advantages and drawbacks regarding the dynamic range, stability, pattern noise generation and sensitivity to the imperfections of the circuitry have been pointed out.

A compilation of reported A/D converter ICs has been made that define their state of the art. The efficiency of these designs has been evaluated using a figure of merit (*FOM*) that combines the effective resolution, digital output rate and power consumption. The results show that only 14 modulators with a *FOM* smaller than 10pJ have been published in the period 1991-1997. The modulators designed in this book compare favorably with respect to the state-of-the-art modulators. This demonstrates that the optimization techniques used are capable of generating small-*FOM* $\Sigma\Delta$ modulators; that is, with adjusted power dissipation regarding the specifications.

Chapter **3**

Modeling of error mechanisms in Sigma-Delta modulators

3.1   INTRODUCTION

Expressions obtained in the previous chapter for different modulator architectures take into account only quantization noise. However, even though it is usually accepted that ΣΔ conversion is intrinsically less sensitive to the building block non-idealities than other data conversion techniques [Cand92], it is necessary to take into account the impact of the error mechanisms associated with such non-idealities for electrical implementations. The importance of these increases when the specifications of the modulator are demanding because they can become the dominant error sources [Bose88a][Dias92b][Yuka87]. Analyzing the basic block non-idealities is needed with two well differentiated objectives: on the one hand, the obtainment of behavioral models that support a fast and precise time-domain simulation; on the other, the attainment of approximate equations that, in closed form, express the power of the error caused by each non-ideality, as a function of itself and other design variables. This chapter is devoted to the obtainment of such equations, which will be combined with statistic optimization methods for the automatic synthesis of ΣΔ modulators.

The non-idealities which degrade the behavior of ΣΔ modulators can be grouped in two categories:

a) those which produce changes in the signal transfer function (*STF*) and in the quantization noise transfer function (*NTF*). As will be seen, this type of error strongly depends on the architecture of the modulator.

b) those whose effect can be modeled as an error source at the integrator input, and that, hence, do not alter the position of the poles of its transfer function. Traditionally, the approximate analysis of these non-idealities is limited to considering the contribution of the first integrator of the chain.

This is possible because the errors of this type generated in the first integrator are added directly to the input signal and, thus, do not become attenuated in the base band. The contributions of the other integrators to the in-band error power are attenuated by different powers of the oversampling ratio. This way, though the following study is based on a second-order architecture, the results will be applicable to other modulators.

Among the non-idealities belonging to the first group, analyzed in Section 3.2, we have the finite DC-gain of the amplifiers and the mismatching in integrator weights. The following Section 3.3 and 3.4 are devoted to the study of the influence of the integrator dynamic response and to thermal noise, respectively. In the last section, a miscellany of non-ideal effects is included that play a secondary role but that should be considered in the design to ensure its feasibility; for instance, non-linearity of the amplifier DC-gain, capacitor non-linearity and jitter noise.

Intentionally, the analysis of the influence of the quantizer and D/A converter error is not covered in this chapter. Among these errors, the non-linearity is the one which presents a greater impact in the modulator performance. However, in many architectures, the quantizer is reduced to a simple comparator whose linearity is assured by construction. Other comparison errors, such as offset and hysteresis, will be analyzed in the Chapter 4. The effect of the non-linearity will be taken into account in Chapter 7, in the context of multi-bit quantization $\Sigma\Delta$ modulators.

3.2   NON-IDEALITIES THAT AFFECT THE TRANSFER FUNCTION OF THE QUANTIZATION NOISE

The expressions for the in-band quantization noise power in the previous chapter were obtained supposing that the integrator, used as low-pass filter, was ideal. However, in practice, the transfer function of the integrator is different from (1.8) due to the influence of the non-idealities of the electrical implementation. This fact provokes changes in the transfer functions of the signal and quantization noise $(STF(z)$ and $NTF(z)$, respectively), and degrades the modulator performance. This section covers two of the above mentioned non-idealities for switched-capacitor circuits: finite DC-gain of the amplifiers and mismatching in capacitor ratios implementing the integrator weights. As will be seen, because they introduce changes in the transfer functions of the modulator, the influence of both non-idealities depends to a large extent on the architecture selected.

3.2.1 Amplifier DC-gain

At DC, where $z = e^{j2\pi(f/f_S)} \to 1$, the ideal transfer function of the integrator (2.8) assumes that its gain is infinite, a characteristic impossible to obtain in practice [Alle87]. Let us consider, for example, the SC integrator of Fig. 3.1 [Mart79], where the amplifier has been modeled with a voltage-controlled voltage source with gain $A_V \gg 1$. The difference equation for said circuit is

$$v_{o,n} \cong \frac{A_V g_i}{A_V + 1 + g_i} v_{i,n-1} + \frac{(A_V + 1)}{A_V + 1 + g_i} v_{o,n-1} \tag{3.1}$$

where $g_i = C_1/C_2$ is the weight or integrator gain. In the Z-domain we have

$$\frac{V_o(z)}{V_i(z)} = g_i z^{-1} \frac{\dfrac{A_V}{A_V + 1 + g_i}}{1 - \dfrac{A_V + 1}{A_V + 1 + g_i} z^{-1}} \cong \frac{g_i z^{-1}}{1 - (1 - \mu)z^{-1}} \qquad A_V \gg 1 \tag{3.2}$$

with $\mu = g_i/A_V$.

The result is known as lossy integration because only a part of the integrator output in the previous period is added to the new input. Including the transfer function of the lossy integrator in (2.6) yields:

$$STF(z, \mu) \cong \frac{1}{1 + \mu} \cong 1 - \mu$$

$$\mu \ll 1 \qquad z \to 1 \tag{3.3}$$

$$NTF(z, \mu) \cong \frac{1 - z^{-1} + \mu z^{-1}}{1 + \mu} \cong (1 - \mu)(1 - z^{-1}) + \mu z^{-1}$$

So, on the one hand, a small error in the modulator gain is produced and, additionally, a change in the quantization noise shaping function due to a dis-

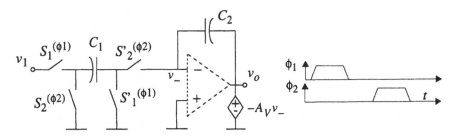

Figure 3.1: SC Integrator and clock phases

placement of the zero from its nominal position in DC. This effect is more explicit in the frequency domain:

$$|NTF(f, \mu)|^2 \cong \mu^2 + (1 - \mu)|NTF(f, 0)|^2 = \mu^2 + 4(1 - \mu)\sin^2\left(\pi\frac{f}{f_S}\right) \qquad (3.4)$$

where f_S is the sampling frequency. Note that this expression is reduced to the ideal case, (2.10), if we made $\mu = 0$. In (3.4) the frequency-independent term is responsible for an increase in the quantization noise power that can be calculated as

$$P_Q(\mu) = \int_{-f_d/2}^{f_d/2} E(f)|NTF(f, \mu)|^2 df \cong \frac{\Delta^2}{12}\left(\frac{\mu^2}{M} + \frac{\pi^2}{3M^3}\right) \qquad \begin{array}{c} M = \dfrac{f_s}{f_d} \gg 1 \\[4pt] \mu \ll 1 \end{array} \qquad (3.5)$$

with M being the oversampling ratio, Δ the separation between two consecutive levels at the D/A converter output, and $f_d = 2f_b$ the Nyquist frequency of the input signal with bandwidth f_b.

Expression (3.5) has a term proportional to the squared leakage factor μ, and inversely proportional to M, which, for high oversampling ratio or small amplifier DC-gain, can dominate the total noise power [Ribn91]. In a similar way, a generic expression can be obtained for the in-band quantization noise power, valid for an arbitrary-order single-loop modulator like that in Fig. 2.10.

$$P_Q(\mu) = \frac{\Delta^2}{12}\left\{\frac{\mu^{2L}}{M} + \sum_{m=1}^{L}\frac{\mu^{2(L-m)}\pi^{2m}}{(2m+1)M^{2m+1}}\frac{L(L-1)...(L-m+1)}{m!}\right\} \qquad (3.6)$$

The number of extra error terms, that is, those that depend on μ, grows with the modulator order. Nevertheless, for usual values of the oversampling ratio and amplifiers DC-gain, the dominant error term for a Lth-order modulator is inversely proportional to M^{2L-1}. Thus, for a 2nd-order modulator, the dominant error term is inversely proportional to M^3, for a third-order one to M^5, etc. The dominant terms in (3.6), are those with $m = L - 1$ (proportional to μ^2) and $m = L$ (ideal term), so that it simplifies to:

$$P_Q(\mu) \cong \frac{\Delta^2}{12}\left[\frac{\mu^2\pi^{2L-2}L}{(2L-1)M^{2L-1}} + \frac{\pi^{2L}}{(2L+1)M^{2L+1}}\right] \qquad (3.7)$$

In order to correctly evaluate the effect of the losses, it is convenient to express (3.7) in relative terms. Thus, the increase in the quantization noise power expressed in dB with respect to an ideal modulator is

$$\Delta P_Q(dB) = 10\log\left(1 + \frac{\mu^2 L}{\pi^2} M^2 \frac{2L+1}{2L-1}\right) \tag{3.8}$$

where it can be observed that the greater the modulator order, the less ideal is its behavior for the same value of μ. Anyway, the increase of the noise power (in dB) presents a logarithmic dependency with the modulator order, which indicates a soft growth.

3.2.1.1 Cascade modulators

We will consider a generic cascade modulator [Long88][Mats87][Uchi88] like that of Fig. 3.2. As stated in the previous chapter, the operation of these modulators is based on the re-modulation in each stage of the quantization noise generated in the previous stage. The quantization noise of the first stage(s) is then digitally canceled, so that the modulator output contains the noise introduced in the last stage, filtered by a shaping function of order equal to the summation of the orders of the stages. Under ideal conditions, the Z-domain output of the modulator in Fig. 3.2 results in:

$$Y(z) = z^{-L_T}X(z) + d(1 - z^{-1})^{L_T}E_N(z) \qquad L_T = L_1 + L_2 + ... + L_N \tag{3.9}$$

with which

$$P_Q = \frac{\Delta_N^2}{12} \cdot d^2 \frac{\pi^{2L_T}}{(2L_T + 1)M^{2L_T + 1}} \tag{3.10}$$

where Δ_N is the separation between levels of the Nth-stage quantizer and d is a scalar larger than unity (2 or 4 are usual values), necessary to avoid the premature overloading of the stages (see Section 2.3.3.2). If we include the effect of the integrator leakage, the result can be approximated by:

$$P_Q(\mu) \cong \frac{\Delta_1^2}{12} \cdot \frac{\mu^2 \pi^{2L_1 - 2}L_1}{(2L_1 - 1)M^{2L_1 - 1}} + \frac{\Delta_N^2}{12} \cdot d^2 \frac{\pi^{2L_T}}{(2L_T + 1)M^{2L_T + 1}} \tag{3.11}$$

where we can observe the presence of the extra noise introduced in the first stage[1]. The incremental form of this expression is,

---

1. In fact, the exact expression includes contributions of all integrators of the modulator, each one filtered with an order equal to the number of precedent integrators in the chain. The same approximation as in (3.7) has been applied, valid for usual ranges of μ and M.

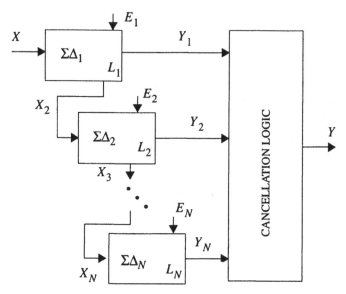

Figure 3.2: Basic scheme of a cascade modulator

$$\Delta P_Q(dB) = 10\log\left[\frac{\Delta_1^2}{\Delta_N^2}\left(1 + \frac{\mu^2 L_1}{\pi^{2(L_T - L_1 + 1)}}\frac{M^{2(L_T - L_1 + 1)}}{d^2}\frac{2L_T + 1}{2L_1 - 1}\right)\right] \qquad (3.12)$$

In this case the increase of the quantization noise power (expressed in dB) grows linearly with the order of the cascade modulator. The exponential dependency on $L_T - L_1 + 1$ in the expression between brackets in (3.12) means that the smaller the order of the first stage, the larger this increase will be. However, to fully exploit the potential of cascade modulators, only unconditionally stable architectures are used as stages; that is, first- and second-order modulators. Then, in order to reduce the effect of the finite gain of the amplifiers, a second-order modulator should be used as the first stage of the cascade modulator ($L_1 = 2$).

Fig. 3.3, obtained by using expressions (2.8) and (2.12), shows the minimum amplifier gain required so that the loss of resolution in respect to the ideal case is only one bit, as a function of the oversampling ratio for several single-loop and cascade architectures. The curves with specified order correspond to single-loop modulators. The rest correspond to cascade modulators, so that each digit expresses the order of a stage; for example, 2-1 refers to a two-stage third-order cascade modulator formed by a second-order stage and a first-order stage, while 1-1-1-1 indicates the cascade of four first-order stages. It is apparent the larger sensitivity of the cascade architectures to the

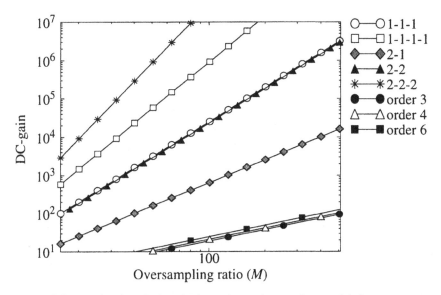

Figure 3.3: Necessary DC-gain in order to lose only one bit in respect to the ideal case

finite gain of the amplifiers respect to their single-loop homonyms. This is the price to pay in order to obviate the stability problems associated with these last. On the other hand, note the difference between the architectures 1-1-1 and 2-1. Both are third order but, as previously mentioned, 1-1-1 presents much larger sensitivity to the integrator leakage.

3.2.2 Capacitor mismatching

In SC circuits, the gain factors are mapped into capacitor ratios [Greg83]. Though these relationships can be obtained with much more precision that the absolute values of the capacitor itself, they are not exempted from error during the fabrication process [Shyu84], so that the gains differ from their nominal values. Again, this type of error modifies the transfer function of the integrator and, consequently, those of the signal and quantization noise, with which its influence depends on the modulator architecture. On the other hand, while the sensitivity of the cascade ΣΔ modulators to the finite amplifier gain can be palliated to a large extent through simple circuit techniques[1], the error due to the mismatching cannot.

---

1. For example, using finite gain insensitive integrators [Greg81][Malo83][Naga85].

Let us consider the second-order $\Sigma\Delta$ modulator in Fig. 3.4. Supposing that each integrator gain presents an error in the form $g_i^* = g_i(1 - \varepsilon_{g_i})$, the transfer functions for the signal and quantization noise result in:

$$|STF(z)| \cong (1 - \varepsilon_{g_1})/(1 - \varepsilon_{g_1'})$$

$$|NTF(z)| \cong \frac{(1 - z^{-1})^2}{(1 - \varepsilon_{g_1'})(1 - \varepsilon_{g_2})} \qquad z \to 1 \tag{3.13}$$

where it has been supposed that $g_1' = g_1$ y $g_2' = 2g_1g_2$. The result is, therefore, an error in the modulator gain and a slight increase in the quantization noise, but the position of the zeroes of the transfer function of the latter is not modified. In agreement with the analysis, the curves obtained by simulation in Fig. 3.5 show that integrator gain standard deviations as large as 1% do not degrade the signal-to-noise ratio, which can be applied to other single-loop modulators. This and other simulation results presented in this chapter have been obtained using ASIDES, a behavioral simulator for $\Sigma\Delta$ modulators whose description will be covered in Chapter 4.

The problem arises when cascade modulators are considered. The digital cancellation of the quantization noise (base of the operation of these modulators) requires that certain relationships among integrator gains and digital coefficients are fulfilled (see Section 2.3.3.2). In SC implementations the mismatching in the capacitor ratios modifies the value of the integrator gains so that such relationships are not met, resulting in incomplete cancellation of the quantization noise from the first stage(s) and, hence, in degradation of the SNR.

For the rest of the calculations we will center on a particular case of the architecture of Fig. 3.2. It is the fourth-order cascade modulator shown in Fig. 3.6 with the structure 2-1-1 [Yin94]. In the previous chapter it was shown that, to obtain the behavior of a fourth-order modulator, the relationships of Table 3.1 had to be fulfilled.

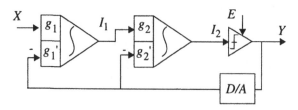

Figure 3.4: Single-loop second-order $\Sigma\Delta$ modulator

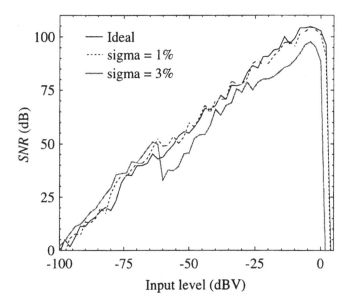

Figure 3.5: Worst-case *SNR* for several integrator weight mismatchings

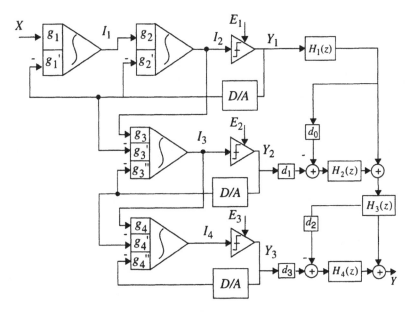

Figure 3.6: Fourth-order 2-1-1 cascade ΣΔ modulator

Table 3.1: Relationships among coefficients for the 2-1-1 modulator

Analog	Digital/Analog	Digital
$g_1' = g_1$	$d_0 = 1 - g_3'/(g_1 g_2 g_3)$	$H_1(z) = z^{-1}$
$g_2' = 2 g_1' g_2$	$d_1 = g_3''/(g_1 g_2 g_3)$	$H_2(z) = (1 - z^{-1})^2$
$g_4' = g_3'' g_4$	$d_2 = \left(1 - \dfrac{g_3'}{g_1 g_2 g_3}\right)\left(1 - \dfrac{g_4'}{g_3'' g_4}\right) \equiv 0$	$H_3(z) = z^{-1}$
	$d_3 = g_4''/(g_1 g_2 g_3 g_4)$	$H_4(z) = (1 - z^{-1})^3$

A calculation similar to that made for the second-order architecture would lead, in this case, given the complexity of the modulator, to cumbersome Z-domain expressions. Instead we will suppose that, as a consequence of the mismatching, the relationships of Table 3.1 are now

$$
\begin{aligned}
d_0 &= [1 - g_3'/(g_1 g_2 g_3)](1 - \varepsilon_0); & g_1' &= g_1(1 - \varepsilon_{g_1'}) \\
d_1 &= [g_3''/(g_1 g_2 g_3)](1 - \varepsilon_1); & g_2' &= 2 g_1' g_2(1 - \varepsilon_{g_2'}) \\
d_2 &= [1 - g_3'/(g_1 g_2 g_3)]\varepsilon_{g_4'}; & g_4' &= g_3'' g_4(1 - \varepsilon_{g_4'}) \\
d_3 &= [g_4''/(g_1 g_2 g_3 g_4)](1 - \varepsilon_3)
\end{aligned}
\tag{3.14}
$$

where each *epsilon* represents the relative error of the digital or analog coefficient and is calculated as

$$
\begin{aligned}
\varepsilon_0 &= \frac{\Delta g_3'}{g_3'} - \frac{\Delta g_1}{g_1} - \frac{\Delta g_2}{g_2} - \frac{\Delta g_3}{g_3} &
\varepsilon_{g_1'} &= \frac{\Delta g_1'}{g_1'} \\
\varepsilon_1 &= \frac{\Delta g_3''}{g_3''} - \frac{\Delta g_1}{g_1} - \frac{\Delta g_2}{g_2} - \frac{\Delta g_3}{g_3} &
\varepsilon_{g_2'} &= \frac{\Delta g_1'}{g_1'} + \frac{\Delta g_2}{g_2} \\
\varepsilon_3 &= \frac{\Delta g_4''}{g_4''} - \frac{\Delta g_1}{g_1} - \frac{\Delta g_2}{g_2} - \frac{\Delta g_3}{g_3} - \frac{\Delta g_4}{g_4} &
\varepsilon_{g_4'} &= \frac{\Delta g_3''}{g_3''} + \frac{\Delta g_4}{g_4}
\end{aligned}
\tag{3.15}
$$

With the coefficients in (3.14) the Z-domain modulator output is

$$
\begin{aligned}
Y|_{2-1-1} \cong{}& X z^{-4} + d_3(1 - \varepsilon_3)(1 - z^{-1})^4 E_3 + d_1(\varepsilon_1 - \varepsilon_3)z^{-1}(1 - z^{-1})^3 E_2 \\
&+ 2(\varepsilon_{g_2'} - \varepsilon_1 - \varepsilon_{g_1'} - \varepsilon_{g_4'})z^{-1}(1 - z^{-1})^3 E_1 + (\varepsilon_1 + \varepsilon_{g_1'})z^{-2}(1 - z^{-1})^2 E_1
\end{aligned}
\tag{3.16}
$$

that is, in addition to the quantization noise of the last stage, E_3, like in (2.28), it contains a portion of the quantization noise of the intermediate stages, E_1 and E_2. Because the order of the shaping functions of these noise contributions is lower than that of the ideal contribution (4 in this case), they can dominate the total in-band noise power. The increase of the quantization noise power expressed in dB, obtained from the expression equivalent to (3.16) in the frequency domain, results

$$\Delta P_Q(dB) = 10\log\left[\left(1 + \frac{9\delta_A^2 M^4}{5 d_3^2 \pi^4} + 4\frac{9\delta_B^2 M^2}{7 d_3^2 \pi^2}\right)\frac{\Delta_1^2}{\Delta_3^2} + \frac{9 d_1^2}{7 d_3^2}\delta_C^2\frac{M^2\Delta_2^2}{\pi^2 \Delta_3^2}\right] \tag{3.17}$$

where

$$\delta_A = \frac{\varepsilon_1 + \varepsilon_{g_1'}}{1 - \varepsilon_3} \qquad \delta_B = \frac{\varepsilon_{g_2'} - \varepsilon_1 - \varepsilon_{g_1'} - \varepsilon_{g_4'}}{1 - \varepsilon_3} \qquad \delta_C = \frac{\varepsilon_1 - \varepsilon_{3'}}{1 - \varepsilon_3} \tag{3.18}$$

For typical values of the oversampling ratio, the term proportional to M^4 dominates in (3.17), so

$$\Delta P_Q(dB) \cong 10\log\left[1 + \frac{9}{5 d_3^2}\left(\frac{\varepsilon_1 + \varepsilon_{g_1'}}{1 - \varepsilon_3}\right)^2 \frac{M^4}{\pi^4}\right] \tag{3.19}$$

In view of (3.15) and (3.19), for this particular architecture, special attention should be paid to the implementation of the coefficients $g_1, g_1', g_2, g_3, g_3'', g_4$ and g_4''.

Singularizing the general study to SC circuits will allows a reduction of the number of independent variables in (3.19). As previously said, in this type of circuit the analog coefficients are implemented through the ratio of two capacitors, that is,

$$g_i = \frac{C_i}{C_i^o} \qquad \Rightarrow \qquad \Delta g_i = g_i\left(\frac{\Delta C_i}{C_i} - \frac{\Delta C_i^o}{C_i^o}\right) \tag{3.20}$$

where Δg_i is a deviation in respect to the nominal value g_i. Assuming that the mismatching in both capacitors is due to statistically independent random processes, the standard deviation of the coefficient value is

$$\sigma_{g_i} = g_i\sqrt{\left(\frac{\sigma_{C_i}}{C_i}\right)^2 + \left(\frac{\sigma_{C_i^o}}{C_i^o}\right)^2} \tag{3.21}$$

where the right member is formed by the standard deviation in the value of the capacitors that implement the coefficient. Such deviations can be calculated as, [Shyu84]

$$\sigma_{nC_u} = C\sqrt{\frac{n^{1/2}K_{le}}{C^{3/2}} + \frac{K_{lo}}{C} + \frac{nK_{ge}}{C} + K_{go}} \tag{3.22}$$

where it has been supposed that the capacitance C has been split into n unitary capacitors with identical value C_u; and the constants K_{le}, K_{ge}, K_{lo}, y K_{go} are related to the local and global effects of the errors of the etching process and the variations in the oxide thickness, respectively. Since these constants are specific for a given fabrication process, the only controllable variables in (3.22) are the value of the unitary capacitor and the number of these that form the capacitor C. If the capacitors C_i and C_i^o in (3.21) are implemented through n_i and m_i unitary capacitor connections, respectively, we obtain

$$\sigma_{g_i} = \frac{n_i}{m_i}\sqrt{\left(\frac{1}{n_i} + \frac{1}{m_i}\right)\left(\frac{K_{le}}{C_u^{3/2}} + \frac{K_{lo}}{C_u}\right) + \frac{2K_{ge}}{C_u} + 2K_{go}} \qquad g_i = \frac{n_i}{m_i} \tag{3.23}$$

Using this expression with the values of the constant estimated for a 0.7μm CMOS technology and 0.25pF unitary capacitors, a gain of 1/2 shows a standard deviation equal to 2.2%. However, using common centroide techniques [Tsiv96] for the implementation of the capacitors allows for the almost total cancellation of the global effects (represented by the constants K_{ge} y K_{go}), reducing the standard deviation to only 0.2%.

The expression (3.23) allows us to calculate the standard deviation of each gain and, consequently, that of the relative errors and combinations of these in (3.15) and (3.18). Nevertheless, note that in (3.15) some coefficient variations are correlated because such coefficients share the integration capacitor. Let us consider, for example, the contributions of g_3 and g_3'' to ε_1 (relative error of d_1). Using (3.20),

$$\sigma^2\left(\frac{\Delta g_3''}{g_3''} - \frac{\Delta g_3}{g_3}\right) = \sigma^2\left(\frac{\Delta C_3''}{C_3''} - \frac{\Delta C_3}{C_3}\right) = \frac{\sigma_{C_3''}^2}{C_3''^2} + \frac{\sigma_{C_3}^2}{C_3^2} \tag{3.24}$$

with which the deviation of the integration capacitor is cancelled. This result seems reasonable because d_1 depends only on the ratio g_3''/g_3 and not on the exact value of C_3^o.

The worst case for each relative error that produces maximum non-cancelled quantization noise, can be estimated as three times the corresponding standard deviation [Papo65].

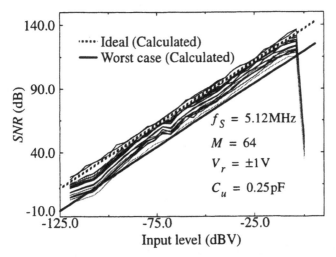

Figure 3.7: *SNR* vs. input level assuming capacitor mismatching

Fig. 3.7 shows a group of *SNR* curves as a function of the input level for the modulator of Fig. 3.6. They resulted from a Monte Carlo simulation considering gaussian distributions for the integrator gains with a mean equal to its nominal value and standard deviation calculated from (3.23). Global effects have been considered compensated ($K_{ge} = K_{go} = 0$). The standard deviation in the *SNR* peak is 4.6%, with a worst case of 115dB, approximately 12dB (2bit) below the ideal case. Note that the worst-case analytical curve (the thick line in Fig. 3.7) fits very well with the worst-case *SNR* obtained through simulation. The expression (3.19) can be used to estimate the minimum capacitor necessary, so that the loss of resolution is only one bit, as shown in Fig. 3.8. Note the strong dependency on the oversampling ratio.

3.3   ERROS DUE TO THE INTEGRATOR DYNAMICS

Incomplete charge transfer constitutes one of the main error sources in SC circuits [Geig82][Lee85][Mart81][Sans87][Teme80]. Particularly for ΣΔ modulators this non-ideal effect is dominated by three characteristics of the operational amplifiers:

a) Finite gain-bandwidth product (*GB*).
b) Slew-rate (*SR*), or limitation of the output voltage change rate.
c) Finite and non-linear DC open-loop gain.

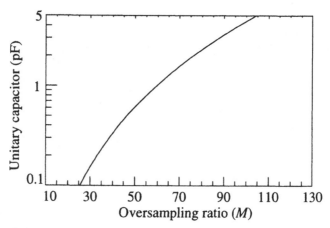

Figure 3.8:   Required unitary capacitor so that the loss of resolution
due to mismatching is smaller than one bit

The charge transfer error mechanisms and their impact on the functionality of the SC integrator (mainly gain error and distortion) have been previously studied by several authors [Lee85][Sans87]. Unfortunately, the
analysis developed for linear systems are only partially applicable to the case
of $\Sigma\Delta$ modulators due to the fact that these last are highly non-linear systems
– the grosser the internal quantization, the more non-linear they are
[Cand92]. Concerning the models developed specifically for $\Sigma\Delta$ modulators,
they are not very accurate because it is supposed, for simplicity, that the
errors caused by the incomplete charge transfer can be considered as white
noise [Dias92b], which is not always true as will be seen later.

We will devote this section to the study of the first two non-idealities presented previously, which define the transient response of the amplifiers. The
influence of the finite DC-gain has been covered in the previous section,
while the distortion caused by its non-linearity is envisaged in Section 3.5.2.

3.3.1   Transient response model

Let us consider the single-branch SC integrator of Fig. 3.1. The situation
under study is shown in Fig. 3.9. It shows the evolution of the integrator output node voltage during the integration phase (switches S_2 and S_2' ON), configuration of Fig. 3.9(b). For the theoretical analysis, we will assume a
single-pole model for the operational amplifier[†1], which is approximately the

---

1. This is a good approximation for a single-stage amplifier with large phase margin
(≥ 60degree) [Gray93].

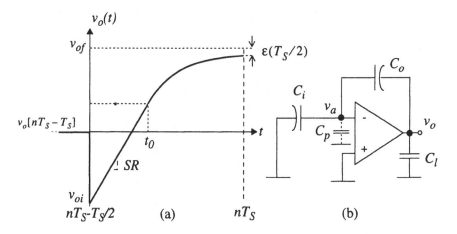

Figure 3.9:   (a) Time-evolution of the integrator output node. (b) Integration phase configuration.

case in many practical ΣΔ modulator designs. With that, consider the equivalent circuit of Fig. 3.10 where, in addition to the input and output parasitics capacitors, a voltage-controlled non-linear current source has been included to model the current saturation of the amplifier.

Returning to Fig. 3.9(a), due to the fact that the amplifier is of high output impedance, the principle of the charge conservation makes the integrator output voltage v_o jump in the opposite direction to that of the final increase, presenting a discontinuity in the instant in which the switches S_2 and S_2' are closed [Sans87]. A similar evolution occurs in the amplifier input node whose voltage v_a experiences a jump, this time departing from zero (supposing offset null), toward negative voltages if the final increase of v_o is positive. After this initial variation, the amplifier transfers the charge stored in C_i

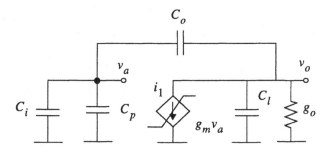

Figure 3.10:   Equivalent circuit during the integration phase

to C_o during the rest of the integration phase, so that $v_a(t)$ approaches zero and $v_o(t)$ to its final value v_{of} imposed by the transfer function of the integrator.

The value of $v_a(t)$ reached immediately after the commutation of the switches is obtained by applying the principle of charge conservation slightly before and after the commutation, resulting in:

$$v_{ai} = -\frac{C_i}{C_{eq}}\left(1 + \frac{C_l}{C_o}\right)v_i, \qquad C_{eq} = C_i + C_p + C_l\left(1 + \frac{C_i + C_p}{C_o}\right) \qquad (3.25)$$

being v_i the voltage stored in C_i. The value of the output voltage in that instant is

$$v_{oi} = v_o[nT_S - T_S] - \frac{C_o}{C_l + C_o}v_{ai} = v_{o,n-1} - \frac{C_i}{C_{eq}}v_i \qquad (3.26)$$

Depending on the absolute value of v_{ai} we will distinguish two cases:

a) Small increases at the input, $|v_{ai}| < I_o/g_m$, where I_o is the maximum current that can be supplied by the amplifier and g_m is its transconductance. In such cases, the operation of the circuit of Fig. 3.10 is linear, that is, the amplifier is not saturated in current, by which:

$$v_o(t) = \frac{v_{o,n-1} + C_i/C_o v_i}{1 + \frac{g_o}{g_m}\left(1 + \frac{C_i + C_p}{C_o}\right)}$$

$$+ \left\{-\frac{v_{o,n-1} + C_i/C_o v_i}{1 + \frac{g_o}{g_m}\left(1 + \frac{C_i + C_p}{C_o}\right)} + v_{oi}\right\} \cdot \exp\left\{-\frac{g_m + g_o\left(1 + \frac{C_i + C_p}{C_o}\right)}{C_{eq}}t\right\} \qquad (3.27)$$

b) However, in practice the amplifier usually starts the integration phase outside of its linear zone; that is, with $|v_{ai}| > I_o/g_m$. Thus, during the first part of the transient response, the controlled current source supplies a constant current source of value I_o. Analyzing the circuit of Fig. 3.10 for such cases obtains

$$v_a(t) = \frac{I_o \operatorname{sgn}(v_i)}{g_o[1 + (C_i + C_p)/C_o]} - \frac{v_{o,n-1} + (C_i/C_o)v_i}{[1 + (C_i + C_p)/C_o]} \tag{3.28}$$

$$+ \left[-\frac{I_o \operatorname{sgn}(v_i)}{g_o \left(1 + \dfrac{C_i + C_p}{C_o}\right)} + \frac{v_{o,n-1} + (C_i/C_o)v_i}{1 + \dfrac{C_i + C_p}{C_o}} + v_{ai} \right] \cdot \exp\left[-g_o \frac{\left(1 + \dfrac{C_i + C_p}{C_o}\right)}{C_{eq}} t \right]$$

$$v_o(t) = v_{o,n-1} + \left(1 + \frac{C_i + C_p}{C_o}\right) v_a(t) + \frac{C_i}{C_o} v_i \tag{3.29}$$

For $g_o \ll 1$, the expression (3.28) can be approximated by

$$v_a(t) = -\frac{C_i}{C_{eq}}\left(1 + \frac{C_l}{C_o}\right)v_i + \frac{I_o \operatorname{sgn}(v_i)}{C_{eq}} t \tag{3.30}$$

The amplifier enters its linear zone when the equality $|v_a(t_0)| = I_o/g_m$ is reached, where t_0 is calculated from (3.30) as

$$t_0 = -\frac{C_{eq}}{g_m} + \frac{C_i}{I_o}|v_i|\left(1 + \frac{C_l}{C_o}\right) \tag{3.31}$$

For $t > t_0$, it gives

$$v_a(t) = -\frac{v_{o,n-1} + (C_i/C_2)v_i}{1 + \dfrac{g_m}{g_o} + \dfrac{C_i + C_p}{C_2}} + \left[\frac{v_{o,n-1} + (C_i/C_2)v_i}{1 + \dfrac{g_m}{g_o} + \dfrac{C_i + C_p}{C_2}} - \frac{I_o}{g_m}\operatorname{sgn}(v_i) \right] \tag{3.32}$$

$$\cdot \exp\left\{ -\frac{g_m + g_o[1 + (C_i + C_p)/C_o]}{C_{eq}}(t - t_0) \right\}$$

$$v_o(t) = v_{o,n-1} + \left(1 + \frac{C_i + C_p}{C_o}\right)v_a(t) + \frac{C_i}{C_o} v_i \tag{3.33}$$

Using (3.27), (3.32) and (3.33) the value in the end of the integration phase (after $T_S/2$) can be calculated:

$$v_o\left(\frac{T_S}{2}\right) \tag{3.34}$$

$$= \begin{cases} v_{o,n-1} + \dfrac{C_i}{C_o}v_i - \left(\dfrac{C_i}{C_o} + \dfrac{C_i}{C_{eq}}\right)v_i \cdot \exp\left(-\dfrac{g_m}{C_{eq}}\dfrac{T_S}{2}\right) & |v_{ai}| \leq \dfrac{I_o}{g_m} \\[4mm] v_{o,n-1} + \dfrac{C_i}{C_o}v_i - \left(1 + \dfrac{C_i+C_p}{C_o}\right)\dfrac{I_o}{g_m}\,\mathrm{sgn}(v_i)\cdot\exp\left[-\dfrac{g_m}{C_{eq}}\left(\dfrac{T_S}{2}-t_0\right)\right] & , |v_{ai}| > \dfrac{I_o}{g_m} \end{cases}$$

where it has been supposed that $g_m \gg g_o$.

Finally, during the following sampling phase, the voltage stored in C_o keeps constant while $v_a(t)$ tends to zero. We will suppose that the output of the integrator is sensed during this phase, so that

$$v_o(T_S) = v_o\left(\frac{T_S}{2}\right) - v_a\left(\frac{T_S}{2}\right) \tag{3.35}$$

$$= \begin{cases} v_{o,n-1} + v_i\dfrac{C_i}{C_o}\left[1 - \dfrac{C_i+C_p}{C_i}\varsigma\cdot\exp\left(-\dfrac{g_m}{C_{eq}}\dfrac{T_S}{2}\right)\right] \\[2mm] \qquad \text{si } |v_i| \leq I_o/(g_m\varsigma) \\[4mm] v_{o,n-1} + v_i\dfrac{C_i}{C_o} - \dfrac{C_i+C_p}{C_o}\dfrac{I_o\,\mathrm{sgn}(v_i)}{g_m}\exp\left[-\dfrac{g_m}{C_{eq}}\left(\dfrac{T_S}{2}-t_0\right)\right] \\[2mm] \qquad \text{si } |v_i| > I_o/(g_m\varsigma) \end{cases}$$

where

$$\varsigma = \frac{C_i}{C_{eq}}\left(1 + \frac{C_l}{C_o}\right) \tag{3.36}$$

In both cases of (3.35), some error is induced in the integrator output voltage which can be modeled by including error terms in the integrator gain as follows:

$$g_i^{nl}(v_i) = \begin{cases} g_i(1 - \beta e); & |v_i| \leq v_L \\ g_i(1 - \beta e^{|v_i|/v_L}); & |v_i| > v_L \end{cases} \tag{3.37}$$

where

$$\beta = \left(1 + \frac{C_p}{C_i}\right)\varsigma \cdot \exp\left(-\frac{g_m}{C_{eq}}\frac{T_S}{2} - 1\right) \qquad v_L = \frac{I_o}{g_m\varsigma} \qquad g_i = \frac{C_i}{C_o} \tag{3.38}$$

Fig. 3.11 is a plot of the expression (3.37). Note that, when the input signal is below the limit v_L, the gain of the integrator does not depend on the input signal; so the error maps into an extra noise at the modulator output that we will call *incomplete settling noise*. On the other hand, in the second case, there exists a dependency between the gain of the integrator and its input, which leads to a non-linear gain (in fact, to obtain (3.37) it has been supposed that such non-linearity is weak) and, consequently, distortion at the modulator output. We will devote the following sections to evaluate separately both error mechanisms.

3.3.2 Incomplete settling noise

First consider that there is no dependency between the gain of the integrator and its input. In such a case, the incomplete settling error is

$$g_i^{nl}(v_i) = g_i(1 - \beta e)$$
$$v_o = v_{oi} + g_i(v_i - \beta e \cdot v_i) \qquad \varepsilon = v_i\left(1 + \frac{C_p}{C_i}\right)\varsigma \cdot \exp\left(\frac{g_m}{C_{eq}}\frac{T_S}{2}\right) \tag{3.39}$$

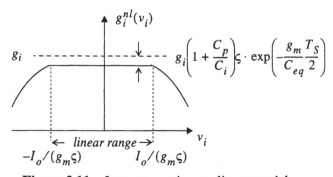

Figure 3.11: Integrator gain non-linear model

Notice that the input of the first integrator consists of the difference between the modulator input and output signals (Fig. 3.4) and that the latter, except for the quantization noise, is a copy of the input. A little coarse but effective approximation involves supposing that the input of the integrator and the settling error, in the case it is linear, are not correlated. So, we will suppose that such an error presents a gaussian distribution in the range $(-\varepsilon_m, \varepsilon_m)$, being ε_m its maximum value. Said value is obtained for the maximum input level of the integrator in which we will assume $v_i|_{max} = V_r$, where V_r is the quantizer half-scale output, which coincides with $\Delta/2$ when the quantizer is a comparator. Thus, the associated power spectral density is constant in the frequency range $(-f_S/2, f_S/2)$ where $f_S = (T_S)^{-1}$, and can be calculated as

$$S_{st}(f) = \frac{\varepsilon_m^2}{9 f_S} = \frac{V_r^2}{9 f_S}\left(1 + \frac{C_p}{C_i}\right)^2 \varsigma^2 \exp\left(-\frac{g_m}{C_{eq}}T_S\right) \tag{3.40}$$

where it has been supposed $\sigma_\varepsilon = \varepsilon_m/3$. Integrating (3.40) in the signal band $(-f_b, f_b)$ results in the error power due the incomplete settling:

$$P_{st} = \frac{V_r^2}{9M}\left(1 + \frac{C_p}{C_i}\right)^2 \varsigma^2 \exp\left(-\frac{g_m}{C_{eq}}T_S\right) \qquad M = \frac{f_S}{2 f_b} \tag{3.41}$$

3.3.3 Slew-rate distortion

As stated, during the integration phase, if the maximum current that the amplifier can supply is exceeded, a dependency of the integrator gain on its input is created which causes distortion. In such a case, for sinusoidal inputs, harmonics are produced whose power can dominate the in-band error power. In this section the extent of these harmonics is evaluated.

It can be demonstrated [Op'T93] that, for a $\Sigma\Delta$ modulator, whose first integrator presents a non-linear gain that we will represent generically as:

$$g_i^{nl}(v) = g_i \sum_n \alpha_n v^{n-1} \qquad \alpha_n \ll \alpha_{n-1} \; ; \; n = 1, 2, 3... \tag{3.42}$$

the amplitude of the n-th harmonic at the output of the modulator results in:

$$A_{H,n} = \alpha_n A^n/(2^{n-1}k_1); \; n = 1, 2, 3, ... \tag{3.43}$$

where A is the amplitude of the input sinusoide. In our case, we will consider

$\alpha_2 = \alpha_4 = 0$ because the gain of the integrator described by (3.37) is symmetrical (Fig. 3.11). Such symmetry, even in the gain curve, is odd in the transfer function, therefore the series expansion of the latter will contain only odd powers. Truncating the series in the fifth term of (3.42) yields

$$g_i^{nl}(v_i) = g_i(\alpha_1 + \alpha_3 v_i^2 + \alpha_5 v_i^4) \tag{3.44}$$

For real amplifiers the maximum supplied output current can be different for positive or negative output swings, making the curve that represents the gain of the integrator non-symmetrical. However, such dissymmetry must be small in a correctly designed amplifier, so that the even-order harmonics will be negligible with respect to odd-order ones.

At this point, in order to evaluate the harmonic distortion we need to determine a curve that, with the form given in (3.44), best fits (3.37) in a given interval. Considering a ΣΔ modulator like that of Fig. 3.4, the signal at the first integrator input is distributed in the interval $(k_1 V_r{-}A,\ k_1 V_r{+}A)$, with $k_1 = g_1'/g_1$ and A equal to the input amplitude. This fact is corroborated by the histogram of the first integrator input for $k_1 = 0.5$, $V_r = 1V$ and $A = 0.3V$ shown in Fig. 3.12.

A solution to this problem involves using non-linear regression, applying the method of minimum square error [Vand84] in such a range. Thus, the calculation of the coefficients in (3.44) is reduced to the solution of the following set of linear equations:

$$\sum_{j=1,3,5} \alpha_j (v_{R,h}^{j+2l} - v_{R,l}^{j+2l})/(j+2l) = B_l \ ; \ l = 0, 1, 2 \tag{3.45}$$

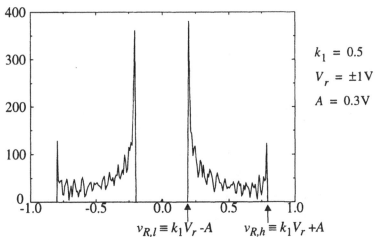

Figure 3.12: Typical first-integrator input hystogram in a ΣΔ modulator

where $v_{R,h}$ y $v_{R,l}$ represent the upper and lower limits of the interval of interest, respectively, and

$$B_l = \int_{v_{R,l}}^{v_{R,h}} g_i^{nl}(v) v^{2l} dv = \int_{v_{R,l}}^{v_L} g_i(1-\beta e) v^{2l} dv + \int_{v_L}^{v_{R,h}} g_i\left[1-\beta exp\left(\frac{v}{v_L}\right)\right] v^{2l} dv \quad (3.46)$$

where it has been supposed that $v_{R,l} < v_L < v_{R,h}$ [1]. In the worst case, when $A = k_1 V_r$ and therefore $v_{R,l} = 0$, the values α_3 and α_5 are:

$$\alpha_3 \cong -105\beta\left[\frac{9}{8}(11e-30\kappa)\frac{v_L^5}{v_{R,h}^7} + \frac{135}{4}\kappa\frac{v_L^4}{v_{R,h}^6} - \frac{1}{8}(7e+114\kappa)\frac{v_L^3}{v_{R,h}^5} + 3\kappa\frac{v_L^2}{v_{R,h}^4} - \frac{1}{4}\kappa\frac{v_L}{v_{R,h}^3}\right]$$

$$\alpha_5 \cong 315\beta\left[\frac{7}{16}(11e-30\kappa)\frac{v_L^5}{v_{R,h}^9} + \frac{105}{8}\kappa\frac{v_L^4}{v_{R,h}^8} - \frac{5}{16}(e+18\kappa)\frac{v_L^3}{v_{R,h}^7} + \frac{5}{4}\kappa\frac{v_L^2}{v_{R,h}^6} - \frac{1}{8}\kappa\frac{v_L}{v_{R,h}^5}\right]$$

$$(3.47)$$

with $\kappa = exp(v_{R,h}/v_L)$; $v_{R,h} \geq v_L$. Applying (3.43), (3.44) and (3.47), the amplitudes of the third and fifth harmonics at the modulator output are:

$$A_3 = |\alpha_3|A^3/(4k_1); \quad A_5 = |\alpha_5|A^5/(16k_1) \quad (3.48)$$

Fig. 3.13 shows the result of a behavioral simulation for a fourth-order cascade architecture. When the output spectrum is compared to the ideal case, a slight increase of the noise in the signal band is observed, clearly

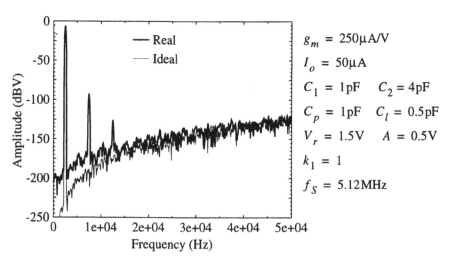

Figure 3.13: Harmonic distortion due to the integrator dynamics

---

1. Note that the validity of the calculations is restricted to the case $v_{R,h} > v_L$; otherwise the settling will be linear for whatever output swing, so that no distortion will be produced.

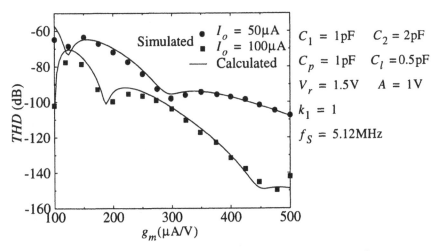

Figure 3.14: *THD* as a function of the amplifier transconductance

dominated by the appearance of the third and fifth harmonics. In Fig. 3.14 the total harmonic distortion (*THD*) calculated using the expressions (3.47) and (3.48) is compared to that obtained through behavioral simulation as a function of the amplifier transconductance for two values of the maximum current. Notice that the curves are not monotonic; that is, for a given value of the maximum current, there are two or more transconductance values that allow the same level of *THD* to be obtained.

3.4   THERMAL NOISE

3.4.1   Noise power spectral density in sampled systems

We will begin obtaining the transfer function for the ideal track-and-hold circuit of Fig. 3.15, where $x(t)$ represents a generic signal. Graphically, the operation of this circuit is shown in Fig. 3.16. Note that the sampled signal,

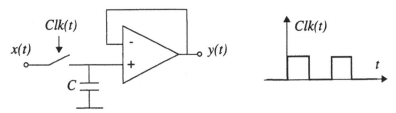

Figure 3.15: Ideal track-and-hold circuit

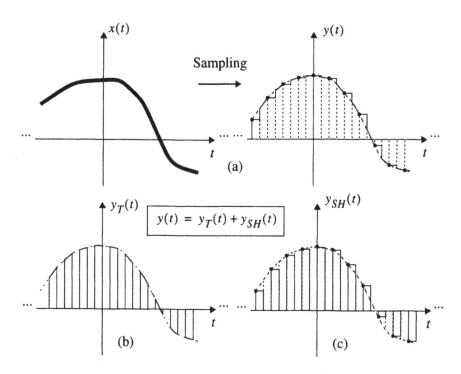

Figure 3.16: (a) Track-and-hold operation: (b) sampled signal during the *tracking* phase, and (c) during the *sample-and-hold* phase.

$y(t)$, can be obtained as the sum of two functions, $y_T(t)$ and $y_{SH}(t)$, corresponding to each operation of the circuit: (a) the tracking operation, when the switch is ON and the voltage in the capacitor follows the input signal, and (b) the sample-and-hold operation that starts in the last instant of the tracking interval and continues while the switch is OFF; during this interval the voltage sampled in the end of the tracking interval is maintained in the capacitor. In other words, the sampled signal is formed by one continuous nature contribution and another of sampled nature, corresponding to the intervals where the switch is ON and OFF, respectively. We will analyze each contribution separately.

3.4.1.1 Tracking model

The tracking operation is modeled in Fig. 3.17, where τ_T/T_S represents the duty-cycle of the clock; that is, τ_T is the interval in which the switch is ON. The signal $c(t)$ in Fig. 3.17 is periodic with period T_S and can be expressed through a series expansion of fasors, as follows:

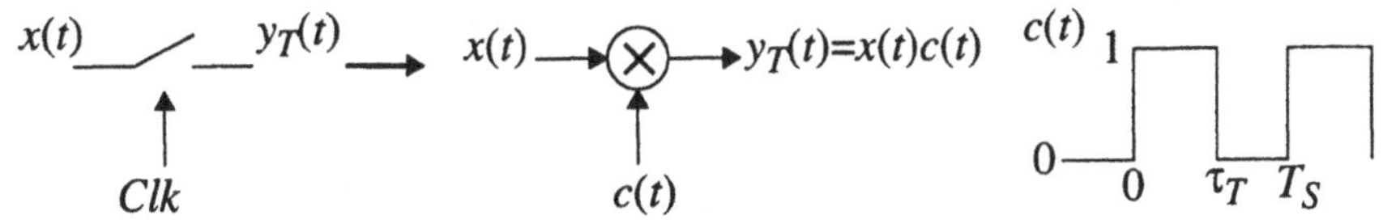

Figure 3.17: Ideal track-and-hold model during the tracking interval

$$c(t) = \sum_{n=-\infty}^{\infty} C_n exp(jn2\pi f_S t) \; ; \quad f_S = \frac{1}{T_S} \tag{3.49}$$

with

$$C_n = \frac{1}{T_S} \int_0^{T_S} c(t) exp(-jn2\pi f_S t) dt \tag{3.50}$$

thus,

$$c(t) = \sum_{n=-\infty}^{\infty} \frac{[1 - exp(-jn2\pi f_S \tau_T)]}{j2\pi n} exp(jn2\pi f_S t) \tag{3.51}$$

The power spectral density for this function is given by

$$S_c(f) = \sum_{n=-\infty}^{\infty} |C_n|^2 \delta(f - nf_S) \tag{3.52}$$

$$= \left(\frac{\tau_T}{T_S}\right)^2 \sum_{n=-\infty}^{\infty} \sin c^2(n\pi f_S \tau_T) \delta(f - nf_S)$$

where

$$\sin c(x) = \begin{cases} 1 & x \equiv 0 \\ \dfrac{\sin(x)}{x}, & x \neq 0 \end{cases} \qquad \delta(x) = \begin{cases} 1 & x \equiv 0 \\ 0 & x \neq 0 \end{cases} \tag{3.53}$$

On the other hand, the Fourier transform of the time-domain product of two functions is given by the convolution product of their transforms, which also applies to the corresponding power spectral densities

$$S_{y_T}(f) = S_x(f) \otimes S_c(f) = \left(\frac{\tau_T}{T_S}\right)^2 \sum_{n=-\infty}^{\infty} \mathrm{sinc}^2(n\pi f_s \tau_T) S_x(f - nf_S) \qquad (3.54)$$

Let us center on the case in which $x(t)$ is a narrow-band noise coming from the filtering of white noise through a low-pass filter as shown in Fig. 3.18. Note that BW_n represents the equivalent bandwidth of the noise, calculated so that it contains the same power as the represented noise but with constant spectral density S_0; that is

$$S_0 \cdot BW_n = \int_{-\infty}^{\infty} S_0 |H_F(f)|^2 df \qquad (3.55)$$

where $H_F(f)$ is the transfer function of the filter. With this, the particular form of the power spectral density of the filtered noise can be obviated in the following calculations. Returning to the expression (3.54), if $f_S \geq BW_n$ the spectra centered in integer multiples of f_S are not overlapped (there is no aliasing). Consequently (3.54) is reduced to

$$S_{y_T}(f) = \left(\frac{\tau_T}{T_S}\right)^2 S_0 \qquad (3.56)$$

On the contrary, if $f_S < BW_n$, it is shown by inspection that the number of bands that are overlapped in the interval $(-f_S/2, f_S/2)$ is BW_n/f_S. In such a case (3.54) can be written as

$$S_{y_T}(f) = \left(\frac{\tau_T}{T_S}\right)^2 S_0 \left[1 + 2 \sum_{n=1}^{\frac{BW_n}{f_S}} \mathrm{sinc}^2(n\pi f_S \tau_T) \right] \qquad (3.57)$$

expression that, when $BW_n \geq 5 f_S$, can be approximated by

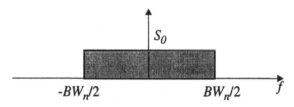

Figure 3.18: Power spectral density of a narrow-band noise

$$S_{y_T}(f) \cong \frac{\tau_T}{T_S}S_0 \qquad (3.58)$$

Thus, the tracking operation reduces the power spectral density of a narrow-band noise even supposing that aliasing is produced. As will be seen, the same does not occur with the sample-and-hold process.

3.4.1.2 Sample-and-hold model

The operation to model is that of Fig. 3.16(c). Note that $y_{SH}(t)$ can be obtained through the following time-domain convolution:

$$y_{SH}(t) = x^*(t) \otimes g_{\tau_{SH}}(t) \Rightarrow Y_{SH}(f) = X^*(f)G_{\tau_{SH}}(f) \qquad (3.59)$$

where the functions $x^*(t)$ and $g_{\tau_{SH}}(t)$ are those shown in Fig. 3.19. Observe that $x^*(t)$ is a sampled version of $x(t)$, that is

$$x^*(t) = \sum_n x[nT_S]\delta(t - nT_S) \qquad (3.60)$$

Taking Fourier transforms

$$X^*(f) = \frac{1}{T_S}\sum_n X(f - nf_S) \qquad (3.61)$$

Furthermore, the Fourier transform of $g_{\tau_{SH}}(t)$ is

$$G_{\tau_{SH}}(f) = \tau_{SH}\operatorname{sinc}(\pi f \tau_{SH}) \qquad (3.62)$$

by which, according to (3.59)

$$S_{y_{SH}}(f) = S_{x^*}(f)\left|G_{\tau_{SH}}(f)\right|^2 \qquad (3.63)$$

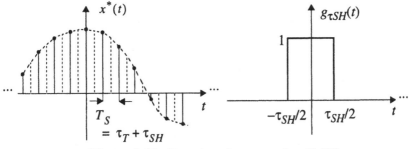

Figure 3.19 Functions in expression (3.59)

$$S_{y_{SH}}(f) = S_{x*}(f)\tau_{SH}^2 \mathrm{sin}c^2(\pi f \tau_{SH}) \tag{3.64}$$

Again, if the power spectral density of $x(t)$ is similar to that shown in Fig. 3.18, we will distinguish two cases:

a) $f_S \geq BW_n$; that is, there is no aliasing, and with (3.61)

$$S_{x*}(f) = \frac{S_0}{T_S^2} \quad ; \quad -\frac{f_S}{2} \leq f \leq \frac{f_S}{2} \tag{3.65}$$

Combining (3.64) and (3.65),

$$S_{y_{SH}}(f) = \left(\frac{\tau_{SH}}{T_S}\right)^2 S_0 \mathrm{sin}c^2(\pi f \tau_{SH}) \cong \left(\frac{\tau_{SH}}{T_S}\right)^2 S_0 \quad ; \quad f \tau_{SH} \ll 1 \tag{3.66}$$

b) $f_S < BW_n$; in such a case

$$S_{x*}(f) = \frac{S_0 BW_n}{T_S^2 f_S} \quad ; \quad -\frac{f_S}{2} \leq f \leq \frac{f_S}{2} \tag{3.67}$$

$$S_{y_{SH}}(f) = \left(\frac{\tau_{SH}}{T_S}\right)^2 S_0 \frac{BW_n}{f_S} \mathrm{sin}c^2(\pi f \tau_{SH}) \cong \left(\frac{\tau_{SH}}{T_S}\right)^2 S_0 \frac{BW_n}{f_S} \tag{3.68}$$

Since the operations of tracking and sample-and-hold are not overlapped, that is $\tau_T = T_S - \tau_{SH}$, we will assume that the noise contributions of both are non-correlated, so that

$$S_y(f) = \begin{cases} S_0\left[\left(\frac{\tau_{SH}}{T_S}\right)^2 \mathrm{sin}c^2(\pi f \tau_{SH}) + \left(1 - \frac{\tau_{SH}}{T_S}\right)^2\right] \quad ; \quad f_S \geq BW_n \\[3mm] S_0\left[\frac{BW_n}{f_S}\left(\frac{\tau_{SH}}{T_S}\right)^2 \mathrm{sin}c^2(\pi f \tau_{SH}) + \left(1 - \frac{\tau_{SH}}{T_S}\right)\right] \quad ; \quad f_S < BW_n \end{cases} \tag{3.69}$$

In view of this expression we can state that, when aliasing is produced, the spectral density of the sampled noise increases in its low-frequency range, with respect to that of the input noise, mainly due to the sample-and-hold operation[1]. This increase is proportional to the relationship between the equivalent bandwidth of the noise and the sampling frequency. This relationship is called the *undersampling ratio*.

---

1. The continuous term will dominate when $\tau_{SH} \to 0$, that is, when the noise is not sampled.

3.4.2 Input-equivalent thermal noise of an SC integrator

In this section we will apply the previous results to calculate the equivalent noise at the input of an SC integrator like that of Fig. 3.20. In this type of architecture the noise generated by the non-zero ON resistance of the switches and that of the amplifier is sampled, together with the signal, at the input capacitor. On the other hand, the need for minimizing the errors in charger transfer recommends a selection of time constants that are small compared to the clock period. This implies a cut-off frequency for the noise several times larger than half the sampling frequency. Thus, the second case of (3.69) applies, by which thermal noise is folded back into the signal band. This phenomenon, that can increase the thermal noise power by a factor of around 100 in a typical design, constitutes the fundamental resolution limit in SC $\Sigma\Delta$ modulators.

Fig. 3.21 shows a simplified model of the SC integrator for the noise analysis. In this model the sources of noise have been substituted by voltage sources of value equal to the mean squared value of the corresponding noise. The series connected ON-resistances of each pair of switches have been substituted by an equivalent resistor whose value has been supposed identical for all branches. In order to simplify the analysis, an infinite gain model has been considered for the amplifier. To perform the analysis we will suppose that all sources are of white noise and are not correlated, which permits the application of the principle of superposition.

3.4.2.1 Noise from switches controlled by ϕ_1

Let us center on Fig. 3.21(a), which corresponds to the sampling phase. During this phase the noise from the switches S_1-S_1' and S_4-S_4' is sampled by

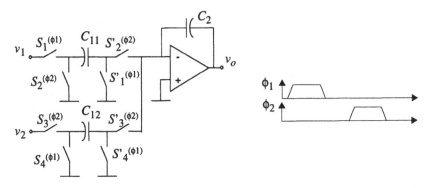

Figure 3.20: Two-branch SC integrator and clock phases

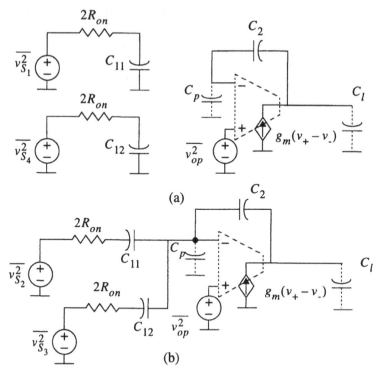

Figure 3.21: Model for the noise analysis of an SC integrator: (a) during the sampling phase; (b) during the integration phase.

the input capacitors C_{11} and C_{12}, respectively. At the same time, these noises are filtered by the low-pass filter formed by the ON resistance of the switches and the capacitor itself, with cut-off frequency several times larger than half the sampling frequency. Applying the second case of (3.69), the noise power spectral densities in both capacitors are

$$S_{C_{11}}(f) = 2 \cdot 2kTR_{on}\left(\frac{1}{2 \cdot 2R_{on}C_{11}f_S}\right)\left[\left(\frac{\tau_{SH}}{T_S}\right)^2 sinc^2(\pi f\tau_{SH}) + \left(1 - \frac{\tau_{SH}}{T_S}\right)\right]$$

$$S_{C_{12}}(f) = 2 \cdot 2kTR_{on}\left(\frac{1}{2 \cdot 2R_{on}C_{12}f_S}\right)\left[\left(\frac{\tau_{SH}}{T_S}\right)^2 sinc^2(\pi f\tau_{SH}) + \left(1 - \frac{\tau_{SH}}{T_S}\right)\right]$$

(3.70)

where it has been made $BW_n = 1/(2 \cdot 2R_{on}C_{1j})$ with $j = 1, 2^{\dagger 1}$ and

---

1. The equivalent bandwidth for a first-order low-pass filter can be calculated applying (3.55) and it results in $BW_n = \pi f_p = \omega_p/2$, where f_p is the frequency of the pole.

$S_0 = 2 \cdot 2kTR_{on}$, being k the Boltzman constant and T the absolute temperature. These expressions can be simplified for the low-frequency range, $f\tau_{SH} \to 0$, resulting in:

$$S_{C_{11}}(f) = \frac{kT}{f_s C_{11}}\left(\frac{\tau_{SH}}{T_S}\right)^2 \qquad S_{C_{12}}(f) = \frac{kT}{f_s C_{12}}\left(\frac{\tau_{SH}}{T_S}\right)^2 \qquad (3.71)$$

once the continuous or tracking part has been neglected. Note that this result, usually called kT/C noise, is independent of the value of the ON-resistance of the switches (keeping in mind that it must be sufficiently small as to permit the correct settling of the signals), because the resistance acts as a filtering element of the noise generated by itself.

Let us suppose that the integrator of Fig. 3.20 is used as the input stage of a $\Sigma\Delta$ modulator, so that the input of the modulator is connected to v_1, while the feedback signal from the D/A converter (see Fig. 3.4) is connected to v_2. In order to calculate the noise power in the modulator, the noise spectral density of the integrator, given in (3.71), must be referred to the modulator input that, with the previous arrangement, coincides with v_1. The noise sampled in C_{11} is added directly to the input signal. With respect to that in C_{12}, note that during the integration phase both capacitors transfer a charge to the feedback capacitor that is proportional to the stored voltage. Thus, the equivalent voltage in C_{11} will be

$$v_{eq,\, C_{11}} = v_{C_{11}} + (C_{12}/C_{11})v_{C_{12}} \qquad (3.72)$$

where the first term is the voltage in C_{11}, while the second represents the voltage, again in C_{11}, that would provoke a charge transfer equal to that of C_{12}. Note that (3.72) refers to the instantaneous value of the voltages in both capacitors in the end of the sampling phase, but if we suppose that the noises generated by different branch resistances are not correlated, a similar expression in terms of power is obtained:

$$\overline{v_{eq,\, C_{11}}^2} = \overline{v_{C_{11}}^2} + (C_{12}/C_{11})^2\overline{v_{C_{12}}^2} \qquad (3.73)$$

Considering a noise with constant power spectral density (to be precise, white noise filtered by a low-pass filter), a similar relationship to (3.73) is applicable to the power spectral density, so

$$S_{eq,\, C_{11}}(f) = \left(1 + \frac{C_{12}}{C_{11}}\right)\frac{kT}{f_s C_{11}}\left(\frac{\tau_{SH}}{T_S}\right)^2 \qquad (3.74)$$

3.4.2.2 Amplifier noise

In Fig. 3.21, the equivalent noise at the input of the amplifier, v_{op}, is sampled by the parallel connected capacitors C_{11} and C_{12} during the integration phase. It is useful to consider the Thévenin equivalent of the circuit of Fig. 3.21(b) that is shown in Fig. 3.22, where the Thévenin parameters are

$$Z_T = \frac{C_1 + C_2}{C_2} \frac{1/g_m}{1 + s/s_p}; \quad v_T = \frac{v_{op}}{1 + s/s_p}; \quad s_p = \frac{g_m C_2}{C_1 C_p + C_1 C_2 + C_p C_2} \quad (3.75)$$

With this, v_C (Fig. 3.22) is given in the Laplace domain by

$$v_C \cong \frac{v_{op}}{1 + s/s'_p} \quad s'_p = \frac{g_m}{C_p + (C_{11} + C_{12})(1 + g_m R_{eq})} \quad R_{eq} = 2R_{on} \quad (3.76)$$

where, for the sake of simplicity, the high-order poles and the value of C_l have been neglected. In this way, we have what is necessary to apply (3.69) and

$$S_{C_{11}}(f) = N_0^2 \frac{g_m}{2 f_s C_i} \left(\frac{\tau_{SH}}{T_s} \right)^2 \quad C_i = C_p + C_{11} + C_{12} \quad (3.77)$$

with an equivalent bandwidth $BW_n = g_m/(2C_i)$. N_0 represents the spectral density of the amplifier noise referred to its input (in $V_{rms}/\sqrt{Hz}$) and the previous approximations for the low-frequency range and negligible continuous contribution have been assumed. Note that to obtain (3.77) it has been supposed that the product $g_m R_{on}$ is small, which amounts to saying that the RC time constant is much smaller than the integrator time constant, as usual.

A similar expression to (2.77) is obtained for the spectral density of noise in the capacitor C_{12}, so that

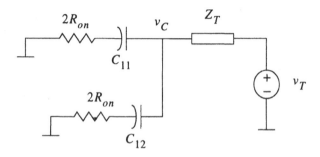

Figure 3.22: Thevenin's equivalent of the circuit in Fig. 3.21(b)

$$S_{eq, C_{11}}(f) = N_0^2 \left(1 + \frac{C_{12}^2}{C_{11}^2}\right) \frac{g_m}{2f_s C_i} \left(\frac{\tau_{SH}}{T_S}\right)^2 \tag{3.78}$$

Note the presence of the input-referred amplifier noise spectral density, N_0. In CMOS implementations [Gray93] N_0 has two main contributions: a flicker noise component, whose spectral density is, in a first approximation, inversely proportional to the frequency, and a white noise component, of thermal origin, whose spectral density is constant. Both noise components are subjected to the sampling operation. However, since the flicker noise density quickly decreases over frequency, the effect of its folding, even for large undersampling ratio, is negligible in comparison with that of the thermal noise. Even more, when the undersampling ratio is increased, all the flicker noise power in the low-frequency region is masked by the folded back white noise. Supposing, therefore, that the flicker component of the amplifier input-referred noise spectral density is negligible, (3.78) can be slightly modified to obtain a relationship more similar to (3.74) as follows:

$$N_0^2 \cong \frac{4kT}{3g_m}$$

$$S_{eq, C_{11}}(f) = \left(1 + \frac{C_{12}^2}{C_{11}^2}\right) \frac{2kT}{3f_s C_i} \left(\frac{\tau_{SH}}{T_S}\right)^2 \tag{3.79}$$

where N_0^2 has been made equal to the spectral density of thermal noise at the input of an amplifier [Gray93]. Note that both expressions, (3.74) and (3.79), can be of the same order of magnitude.

3.4.2.3 Noise from switches controlled by ϕ_2

The treatment of the noise contributions of switches S_2-S_2' and S_3-S_3' is the same as that of the amplifier. The frequency response is dominated by the pole given in (3.76), by which

$$S_{C_{11}}(f) = (2 \cdot 2kTR_{on}) \frac{g_m}{2f_s C_i} \left(\frac{\tau_T}{T_S}\right)^2 \cong \frac{2kTg_m R_{on}}{f_s C_i} \left(\frac{\tau_T}{T_S}\right)^2 \tag{3.80}$$

where it is considered τ_T instead of τ_{SH} because ϕ_1 and ϕ_2 are non-overlapped phases. Again,

$$S_{eq,\,C_{11}}(f) = \left(1 + \frac{C_{12}^2}{C_{11}^2}\right)\frac{2kTg_mR_{on}}{f_SC_i}\left(\frac{\tau_T}{T_S}\right)^2 \tag{3.81}$$

Compare this last expression to (3.74) and (3.79). As stated previously, assuming that $g_mR_{on} \ll 1$ (0.1 or less is a usual value) and that the sampling and integration intervals are of comparable duration, ignoring the noise contributions of switches S_2-S_2' and S_3-S_3' generates only 10% of error. Supposing that all the noises are not correlated, the spectral density of thermal noise can be approximated by

$$
\begin{aligned}
S_{eq,\,C_{11}}(f) \cong \left(\frac{\tau_{SH}}{T_S}\right)^2 & \left[\left(1 + \frac{C_{12}}{C_{11}}\right)\frac{kT}{f_SC_{11}} + \left(1 + \frac{C_{12}^2}{C_{11}^2}\right)\frac{2kT}{3f_SC_i}\right] \\
& + \left(\frac{\tau_T}{T_S}\right)^2\left(1 + \frac{C_{12}^2}{C_{11}^2}\right)\frac{2kTg_mR_{on}}{f_SC_i}
\end{aligned}
\tag{3.82}
$$

3.4.3 Thermal noise in SC $\Sigma\Delta$ modulators

Thermal noise in integrators that form a SC $\Sigma\Delta$ modulator is translated to the output, degrading the performance. However, as for the dynamic errors, only the thermal noise of the first integrator in the chain will be taken into account; since it is added directly to the input signal, it appears with no filtering in the output spectrum. The contributions to the in-band thermal noise power of the rest of the integrators are attenuated by different powers of the oversampling ratio, depending on the position of the integrator and, in general, they can be neglected. The total power of thermal noise after decimation is calculated by integrating $S_{eq,\,C_{11}}(f)$, given by (3.82), in the signal band $(-f_b, f_b)$,

$$P_{th} = \int_{-f_b}^{f_b} S_{eq,\,C_{11}}(f)df \cong \left(1 + \frac{C_{12}}{C_{11}}\right)\frac{kT}{4MC_{11}} + \left(1 + \frac{C_{12}^2}{C_{11}^2}\right)\left(\frac{kT}{6MC_i} + \frac{kTg_mR_{on}}{2MC_i}\right) \tag{3.83}$$

where it has been supposed that the sampling and integration intervals are of identical duration and $M = f_S/(2f_b)$.

At this point we return to the second-order modulator of Fig. 3.4 and particularize for the following practical cases:

a) $g_1 = g_1$'. Note that a two-branch integrator is needed only when both weights are different. Thus, when the input and feedback weights of the first integrator are identical, which means that the loop gain of the modulator is 1, a single-branch integrator can be used and (2.83) results in:

$$P_{th} \cong \frac{kT}{4MC_{11}} + \frac{kT}{2MC_i}\left(\frac{1}{3} + g_m R_{on}\right) \qquad C_i = C_{11} + C_p \qquad (3.84)$$

b) $g_1 \gg g_1'$. This is equivalent to saying that the modulator presents a high loop gain, which can be of interest when the input level of the modulator is very low as, for example, when such input comes from a microsensor. In such a case we can neglect the ratio C_{12}^2/C_{11}^2 in (3.83)

$$P_{th} \cong (1+k_1)\frac{kT}{4MC_{11}} + \frac{kT}{2MC_i}\left(\frac{1}{3} + g_m R_{on}\right) \qquad (3.85)$$

where k_1 represents the inverse of the modulator loop gain

$$\frac{1}{k_1} = \frac{g_1}{g_1'} = \frac{C_{11}}{C_{12}} \qquad (3.86)$$

Note that all previous expressions for the thermal noise power should be multiplied by 2 when referred to fully-differential architectures. With those circuits we can grossly double the signal range to obtain four times more signal power. As thermal noise is just doubled, an increase of 3dB in dynamic range (half a bit) is expected using fully-differential circuitry, which constitutes one of the advantages of such a technique and justifies its wide use.

3.5   OTHER NOISE AND DISTORTION MECHANISMS

Until now, the effects of the more important non-idealities and, hence, the first to take into account when designers face the synthesis of a ΣΔ modulator, have been studied. However, there are other mechanisms of error, more intimately related to the physical implementation, which, depending on the specifications on the converter, may become the dominant error source. Some of them are analyzed in this section.

3.5.1   Harmonic distortion due to the capacitor non-linearity

The imperfections of the capacitors, used as memory elements in SC circuits, influence in a critical way their operation. An example of this was given in Section 3.2.2 where the impact of capacitor ratio mismatching in the performance of cascade $\Sigma\Delta$ modulators was analyzed. On the other hand, the dependency of the value of the capacitance on the stored voltage; that is, the presence of non-linearity, provokes an error in the charge transfer that, as will be seen, is a source of distortion at the modulator output.

Consider the SC integrator of Fig. 3.23(a) where we will suppose that the capacitors are non-linear with a polynomial dependency between the capacitance and the stored voltage v,

$$C(v) = C^o(1 + \alpha v + \beta v^2 + ...) \tag{3.87}$$

where C^o represents the capacitance when the capacitor is uncharged and $\alpha, \beta, ...$ are the non-linear coefficients. The value of these coefficients, expressed in p.p.m./V, p.p.m./V^2, etc., depend on the technique used to implement the capacitor. In particular, for a capacitor implemented with two polysilicon layers, the first non-linear coefficient dominates clearly over the rest. In such a case, we can truncate (3.87) in the linear term, resulting in:

$$C(v) = C^o(1 + \alpha v) \tag{3.88}$$

In the integrator of Fig. 3.23(a), with $v_2 = 0$ and considering an ideal operational amplifier, a differential charge element stored in the capacitor C_1 during phase ϕ_1 is transferred during phase ϕ_2 to the capacitor C_2:

$$dq_1 = dq_2 \; ; \qquad C_1^o(1 + \alpha v)dv = C_2^o(1 + \alpha v)dv$$

Integrating this expression between the initial and final values of the voltage in each capacitor after a clock cycle, yields the following incremental relationship:

$$\int_0^{v_{1,n-1}} C_1^o(1 + \alpha v)dv = \int_{v_{n-1}}^{v_n} C_2^o(1 + \alpha v)dv \tag{3.89}$$

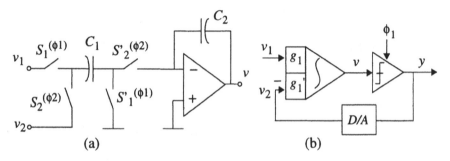

(a)   (b)

Figure 3.23: (a) SC Integrator. (b) First-order $\Sigma\Delta$ Modulator.

$$C_2^o\left(v_n + \frac{\alpha}{2}v_n^2\right) = C_1^0\left(v_{1,n-1} + \frac{\alpha}{2}v_{1,n-1}^2\right) + C_2^o\left(v_{n-1} + \frac{\alpha}{2}v_{n-1}^2\right) \tag{3.90}$$

If the integrator is used in a $\Sigma\Delta$ modulator like that of Fig. 3.23(b), the finite difference equation that represents the integrator operation is

$$C_2^o\left(v_n + \frac{\alpha}{2}v_n^2\right) \tag{3.91}$$

$$= C_1^0\left(v_{1,n-1} + \frac{\alpha}{2}v_{1,n-1}^2\right) - y_n C_1^0\left(V_r + \frac{\alpha}{2}V_r^2\right) + C_2^o\left(v_{n-1} + \frac{\alpha}{2}v_{n-1}^2\right)$$

Note that during phase ϕ_2 the capacitor is charged with constant voltage, V_r, equal to the output of the single-bit D/A converter; because of that, its effect can be obviated for distortion calculations. Reorganizing terms in the previous expression,

$$v_n = v_{n-1} + g_1\left[v_{1,n-1}\left(1 + \frac{\alpha}{2}v_{1,n-1}\right) - v_2\right] + \frac{\alpha}{2}(v_{n-1}^2 - v_n^2) \tag{3.92}$$

where $g_1 = C_1^o/C_2^o$ and $v_2 = y_n V_r$. Including in the second member the ideal expression $v_n = v_{n-1} + g_1(v_{1,n-1} - v_2)$, results in:

$$v_n = v_{n-1} + g_1\left[v_{1,n-1}\left(1 + \frac{\alpha}{2}v_{1,n-1}\right) - v_2\right] - \alpha g_1 v_{n-1}(v_{1,n-1} - v_2) \tag{3.93}$$

$$+ \frac{\alpha}{2}g_1^2(v_{1,n-1} - v_2)^2$$

The last two terms of the second member are proportional to the modulator input minus its output. For sinusoidal inputs, this difference is proportional to $2\pi(f_b/f_S)$, being f_b and f_S the input and sampling frequency, respectively. In $\Sigma\Delta$ modulators $f_b \ll f_S$ with which $2\pi(f_b/f_S) \ll 1$ and such terms can be ignored. The remaining term, proportional to $v_{1,n-1}^2$ generates a second-order harmonic in the modulator output spectrum with amplitude:

$$A_{H,2} = \frac{1}{4}\alpha A^2 \tag{3.94}$$

where A is the input amplitude. The amplitude of this harmonic is doubled for multi-bit modulators.

The analysis of the third-order harmonic distortion due to the capacitance second-oder non-linearity gives results rather more complex, due to the fact that in such a case the contributions of the integration capacitor are not negligible. Nevertheless, an evaluation of this distortion can be easily accomplished through behavioral simulation (see Chapter 4).

3.5.2 Distortion due to the non-linear open-loop DC-gain of the amplifier

Due to a mechanism similar to that just described, the non-linear open-loop gain of the amplifiers introduces error components as harmonic distortion in the modulator output spectrum. The non-linearity of the gain is manifested by its dependency on the amplifier output (or input). In practice, all the amplifiers present, up to a point, a non-linear gain because the transition between the linear and saturation output region is gradual [Gray93]. Fig. 3.24 shows the measured transfer open-loop characteristic measured for an amplifier and the slope of such a curve as a function of the output voltage. Observe that the gain presents a maximum in the center of the scale and decreases as the output voltage approaches the end of the linear region, to fall abruptly once it has surpassed the limit between that and the saturation region.

Considering a finite gain model for the amplifier, the difference equation that describes the operation of the first integrator in the modulator of Fig. 3.4 results in:

$$v_n \cong \frac{A_V v_{i,n-1}}{A_V + 1 + g_1(1 + k_1)} + \frac{(A_V + 1)v_{n-1}}{A_V + 1 + g_1(1 + k_1)} \qquad k_1 = \frac{g_1'}{g_1} \qquad (3.95)$$

an expression similar to (3.1), where it has been considered a two-branch integrator like that of Fig. 3.25, being $v_i \equiv g_1(x \pm k_1 V_{ref})$ the input, and v the output of the integrator.

We will suppose that the amplifier in the first integrator presents an open-

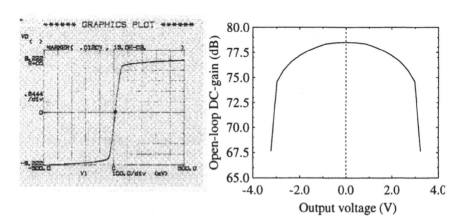

Figure 3.24: (a) Measured DC transfer curve of a CMOS amplifier. (b) Open-loop DC-gain as a function of the output voltage.

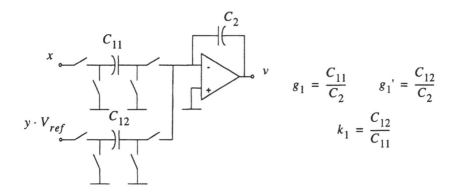

Figure 3.25: Two-branch SC integrator

loop gain whose dependency on the output voltage can be approximated by a polynomial as follows

$$A_V = A_0(1 + \gamma_1 v + \gamma_2 v^2 + ...)$$ (3.96)

Due to the symmetry of the experimental curve in Fig. 3.24(b), the second-order non-linear coefficient is typically negative and of a module quite larger than that of the first order. Considering that A_0 tends to infinite and that γ_1 and γ_2 tend to zero, (3.95) can be approximated by

$$v_n \cong v_{n-1} + g_1 \left[v_i - \frac{v_i + (1 + k_1)v_n}{A_0} + \gamma_1 v_n \frac{v_i + (1 + k_1)v_n}{A_0} + \gamma_2 v_n^2 \frac{v_i + (1 + k_1)v_n}{A_0} \right]$$ (3.97)

where, to simplify, $v_{i,n-1}$ has been replaced by v_i. Thus, the analysis of an SC integrator with a non-linear gain amplifier, as in (3.96), can be accomplished considering an ideal integrator whose input is equal to the expression in brackets in (3.97). The equivalent distortion at the integrator input can be estimated by analyzing the harmonics of such an expression. To do so, we will suppose that the input and the output of the integrator are approximated by their first harmonic, so that

$$v_i \cong V_i \sin(2\pi f_b n T_s) \qquad v_o \cong V_o \cos(2\pi f_b n T_s)$$ (3.98)

where f_b is the frequency of the input and T_S is the sampling period. Substituting these expressions in (3.97) and performing a Fourier series expansion of the term in brackets, the amplitudes of the second and third harmonics referred to the integrator input are:

$$A_{H,2} = \frac{|\gamma_1|}{2A_0}V_o\sqrt{V_i^2 + (1 + k_1)V_o^2} \qquad A_{H,3} = \frac{|\gamma_2|}{4A_0}V_o^2\sqrt{V_i^2 + (1 + k_1)V_o^2} \quad (3.99)$$

The Z-domain equations that describe the behavior of a $\Sigma\Delta$ modulator like that of Fig. 3.4 are:

$$Y(z) = \frac{X(z)z^{-2}}{k_1} + E(z)(1 - z^{-1})^2$$

$$V_i(z) = X - k_1Y(z) = X(z)(1 - z^{-2}) + k_1E(z)(1 - z^{-1})^2 \qquad (3.100)$$

$$V_o(z) = \frac{k_2}{k_1}X(z)z^{-1} + E(z)z^{-1}(1 - z^{-1}) \; ; \; k_1 = \frac{g_1'}{g_1} \qquad k_2 = \frac{g_2'}{g_2}$$

where $X(z)$ and $Y(z)$ represent the modulator input and output respectively, and supposing $k_2 = 2g_1'$. The last two expressions in (3.100) are the Z-transform of the first integrator input and output signals, respectively. According to that, obviating the quantization noise, and taking into account that $2\pi f_b T_S \ll 1$, or equivalently, that z tends to one in (3.100), yields

$$V_i \cong 4\pi f_b T_s A \ll V_o = \frac{k_2}{k_1}A \qquad (3.101)$$

and with (3.99), the amplitudes of the second and third harmonics at the modulator output can be approximated by

$$A_{H,2} = \frac{|\gamma_1|(1 + k_1)k_2^2}{2A_0}\frac{A^2}{k_1^3} \qquad A_{H,3} = \frac{|\gamma_2|(1 + k_1)k_2^3}{4A_0}\frac{A^3}{k_1^4} \qquad (3.102)$$

Note that, if a single-branch integrator can be used, so that the same capacitor is used to sample the input signal and the reference voltage, that is $g_1' = g_1$, k_1 disappears in the previous expressions and a reduction is obtained in the harmonic distortion:

$$A_{H,2} = \frac{|\gamma_1|}{2A_0}k_2^2A^2 \qquad A_{H,3} = \frac{|\gamma_2|}{4A_0}k_2^3A^3 \qquad (3.103)$$

Fig. 3.26 shows the total harmonic distortion (*THD*) obtained through behavioral simulation for a fourth-order cascade architecture (Fig. 3.6). The *THD* is represented as a function of the second-order non-linearity for four

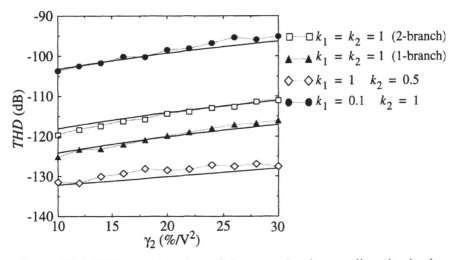

Figure 3.26: *THD* as a function of the second-order non-linearity in the amplifier open-loop DC-gain

sets of values of k_1 y k_2, keeping constant γ_1 equal to 2.5%/V. Observe that the calculated data fit very well with the simulations.

3.5.3 Jitter noise

In practice, the sampling period is not constant but presents variations in its nominal value. This is due to certain intrinsic uncertainties in the time in which clock transitions occur, known as jitter [Taka91]. The result is a non-uniform sampling, responsible for extra noise at the modulator output [Bose88b] that can be estimated as follows.

The error induced when a sinusoidal signal of amplitude A and frequency f_b is sampled in a time instant that does not coincide with a multiple of the sampling period is

$$x(nT + \delta) - x(nT) \cong 2\pi f_b A \delta \cos(2\pi f_b nT_S) \tag{3.104}$$

where δ represents the error in the sampling instant. Supposing that this error has a Gaussian distribution with standard deviation σ_t and mean equal to zero, its power spectral density results in:

$$S_J(f) = \frac{A^2(2\pi f_b \sigma_t)^2}{2} \frac{1}{f_S} \tag{3.105}$$

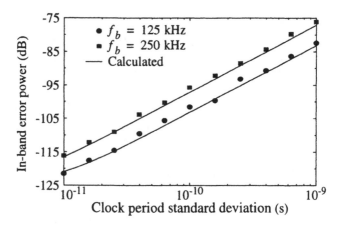

Figure 3.27: In-band jitter noise vs. clock period standard deviation

where f_S is the sampling frequency. Integrating (3.105) in the signal band yields the jitter noise power,

$$P_J = \frac{A^2}{2}\frac{(2\pi f_b \sigma_t)^2}{M} \tag{3.106}$$

Fig. 3.27 shows the jitter noise power calculated using (3.106) as a function of the clock period standard deviation for a fourth-order cascade $\Sigma\Delta$ modulator architecture with $M = 64, A = 1\text{V}, f_b = 125$ and 250kHz. A good fitting between such curves and those obtained through behavioral simulation is observed.

Observe that P_J depends not only on the oversampling ratio but also on the absolute value of the frequency of the input signal. Because of that, the jitter noise plays a very important role in medium-high frequency applications, where it may become the dominant error source. The latter is exemplified in Fig. 3.28 which shows the output spectrum for the ideal case and two frequencies of the input signal with $\sigma_t = 0.1\text{ns}$.

SUMMARY

In this chapter we have analyzed the error mechanisms other than quantization which degrade the performance of the $\Sigma\Delta$ modulation-based A/D converters. These errors, that are expressed as extra noise and/or distortion in the signal-band, are caused by several circuitry imperfections that we called *non-idealities*. In spite of the recognized low sensibility of the $\Sigma\Delta$ modulators to such non-ideal behaviors, when the specifications are very demanding, the in-band power of these errors can dominate that from quantization (the only

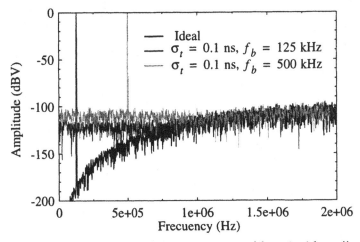

Figure 3.28: Modulator output spectrum with and without jitter

error source present in the simplest theoretical analysis), which justifies the importance of their study and modeling in order to face realistic designs.

Table 3.2 summarizes the non-idealities that to a greater extent degrade the operation of ΣΔ modulators implemented with switched-capacitor circuits. With the exception of the non-idealities that affect the quantizers, whose impact will be analyzed in Chapters 4 and 7, analytical expressions have been obtained that relate each non-ideality to the in-band power of its associated error. Said expressions will be used in Chapter 5 for the automatic design of ΣΔ modulators.

Table 3.2: Main non-idealities that degrade ΣΔ modulator performance

Building block		Non-ideality	Consequence
Integrators	Opamps	DC-gain finite and non-linear	Quantization noise increase, harmonic distortion
		Slew-rate	Harmonic distortion
		Finite GB	Incomplete settling error
		Limited output swing	Overloading
		Thermal noise	White noise
	Switches	Non-zero ON resistance	Settling error, thermal noise
	Capacitors	Non-linearity, mismatching	Quantization noise increase, harmonic distortion
Clock		Jitter	Jitter noise
Comparators		Hysteresis, delay	Quantization noise increase
Multi-bit quantizers		Non-linearity	Harmonic distortion

Chapter 4

Behavioral simulation of Sigma-Delta modulators

4.1 INTRODUCTION

The simulation of electronic circuits is a fundamental step of the design process. This affirmation, applicable to any electronic system, is evident in reference to integrated circuits whose components cannot be modified once manufactured, and for which the calibration is costly and prone to errors. Hence, it is crucial to validate the synthesis through some procedure that allows accurate emulation of the behavior of real circuits.

For basic cell design this process is generally carried out through electrical simulation [Nage75] whose accuracy, provided that precise models for physical devices are available, is high. Regarding the simulation of pure digital circuits, it is relatively easy to increase the degree of abstraction from basic cells, simulated electrically, to digital systems simulated at a logical level, without loss of reliability. This is due on the one hand to the structure of the digital circuits, which allows a clear hierarchization of the systems. On the other, observing a set of clean-looking design rules is enough to guarantee that the operation of the components of the hierarchy is not affected by the rest of the circuitry. It does not occur as such in the analog or mixed-signal systems [Suya90][Dias92a]: on the one hand, it is not always possible to consider the elements of a circuit as isolated entities with independent functionality; on the other, sometimes the hierarchy itself is not well defined.

In such a case, it is clear that, if we want to preserve the quality of the simulation at the same time as the degree of abstraction is increased, we have to continue resorting to the electrical simulation. Unfortunately, though theoretically possible, the electrical simulation of complex systems in a hierarchic form may become unfeasible in terms of computational resources or elapsed time. In particular, for oversampling converters, the extraction of their perfor-

mance involves the analysis of a very larger number of samples at the modulator output [Wolf90][Dias92a], which, in terms of electrical simulation,
means a very long transient analysis. Using this type of analysis, two or three
weeks CPU time would be necessary to estimate the signal-to-noise ratio
from the extracted layout of a typical ΣΔ modulator; supposing that the computers available were able to hold the required memory and calculation
resources.

Accordingly, there is a clear need for other procedures that, with the price
of losing accuracy, can accelerate the simulation. Several alternatives arise,
that in decreasing order concerning precision and computational resources
consumption, can be grouped as follows:

a) Electrical simulation with macromodels of the basic cells.

b) Multi-level simulation.

c) Behavioral simulation, or event-driven simulation.

The first solution, very extended among mixed-signal designers, is still
slow for our purposes, because, though the treatment of the basic cells is simplified, the algorithms to solve the resulting differential equations are numerical and hence excessively costly in CPU time [Meta88].

A recent intermediate approximation involves using multilevel simulators,
from the primitive approximations such as DIANA [Man80] to the most
evolved like ELDO [Anac91], SABER [Saber87] and SWITCAP-2
[Suya90]. Among them, in ELDO and SABER, the critical parts of the system, or those that require greater accuracy in their characterization are simulated numerically, while behavioral models are incorporated to emulate the
rest of the system. It is an interesting solution because the solution of complex equations is performed by the simulator. However, the numerical solution of equations is costly in CPU time (20 hours-CPU to obtain 128ksample
at a typical modulator output [Dias92a]), so that the multi-level simulation is
useful only when the number of time-domain samples, and hence the length
of the transient analysis, does not have to be excessive. On the other hand,
SWITCAP-2 provides the possibility of performing multi-level simulations
with switched-capacitor circuits and arbitrary digital control signals. However, the models for the critical elements of the topology are rather simple
and they do not include some of the most limiting non-idealities for ΣΔ modulators. Furthermore, the CPU time is far from short: 2 hours-CPU to obtain
64ksample in a typical workstation [Dias92a].

The last option (behavioral simulation) tries to bring the simulation of
mixed-signal circuits closer to logic simulation. The so-called event-driven
simulation is applicable to circuits controlled by periodic clock signals,
which form an important group within the analog circuits [Greg83]

[Touma94]. In this procedure, closed expressions for the characteristic input - internal state - output of the basic cells, are used. However, the behavioral simulation leads to reliable results only when the behavioral models include the contour conditions and possible sources of error, whose analysis involves a deep knowledge of the analog circuitry and error mechanisms. In spite of the loss of generality with respect to electrical simulation, the CPU time is drastically reduced. For example, a 45-second CPU time would be enough to obtain 64ksample from a typical modulator [Dias92a].

Such a drastic CPU-time saving has suggested this alternative for shortening the design cycle, so that several tools devoted to the behavioral simulation of ΣΔ converters have been published [Bose88a][Brau90][Dias91a] [Dias91b][Haus86][Nors89][Opal96][Rito88][Wolf90]. The main differences among them are in the number of topologies/basic blocks included and in the precision of their models. These tools also differ in the friendliness of the user interface. Centering on the first point, the tools range from specific architectures treated at a simple difference equation level [Haus86]; that is, from the ideal point of view, or with simplified [Rito88][Nors89] or more elaborated [Wolf90] behavioral models for the basic blocks, to switched-capacitor [Dias91a][Dias91b] or arbitrary [Opal96] architectures with accurate behavioral models.

In any case, none of the simulators which we have had access to (among those the most complete is TOSCA [Dias91a]) covers all the non-idealities limiting the operation of high performance modulators, which has motivated us to develop a tool specifically guided for this purpose. This tool, conceived for the behavioral simulation of SC ΣΔ modulators, is called ASIDES (Advanced Sigma-Delta Simulator). ASIDES is a simulator which focuses more on the physical design than those reported until now, because it includes accurate models for the basic blocks, mainly the integrator, as well as time-domain characterized new basic blocks like multi-bit quantizers and D/A converters. Furthermore, ASIDES includes a wide variety of analysis, from pure time-domain simulation, to the static and dynamic characterization of the converter, including Monte Carlo analysis, parameterized analysis, etc.

In this chapter, after introducing the basic concepts related to the behavioral simulation in Section 4.2, the models included in ASIDES for the response of the SC basic blocks are obtained, taking into account a large number of non-idealities (Section 4.3). Section 4.4 presents the structure of the tool. Finally, Section 4.5 illustrates the use of ASIDES through two practical examples: a second-order modulator − a classical architecture, and a high-performance, more complex architecture − a fourth-order multi-bit cascade modulator.

4.2   BASIC BLOCKS AND BEHAVIORAL SIMULATION

We will begin this section by defining two terms that will be profusely used during the rest of the chapter:

a) Basic block: element of a system whose output is a function of its input and/or its internal state. The law that relates these magnitudes is called the behavioral model of the basic block.

b) Behavioral simulation: a calculation procedure which allows evaluation of the response of a system once the response of its building blocks is known in the form of a behavioral model.

In the first definition it is implicit that, to partition a complex system in basic blocks with independent functionality, one of the following conditions must hold: (a) In said system there is no global feedback loop that could relate the instantaneous output of a basic block to itself. (b) Should such a loop exist, there must be a delay of the signal in order to avoid such dependency. This last condition is fulfilled in SC $\Sigma\Delta$ modulators because they are sampled-data circuits. Consider, for instance, the SC integrator and its clock-phase scheme of Fig. 4.1 [Mart79]. The output of the integrator changes during ϕ_2; that is, when the charge stored in C_i during ϕ_1 is transferred to the capacitor C_o. This means, considering that both phases have the same duration, that the output of the integrator changes according to the value of the input and the output itself half a clock cycle before. If the integrator is followed by another identical, like in a second-order modulator (see Fig. 3.3), the output of the former is not sampled by the latter until ϕ_1 is high again, which supposes one more half-period delay. Thus, the data transmitted from one integrator to the other was obtained with the values of the first integrator input and output in the previous clock cycle. When the loop is closed through a quantizer and a D/A converter, there will not be any instantaneous dependency between the output of the integrator and its input, so that it is possible to consider the integrator as a basic block. This reasoning is extendable to other $\Sigma\Delta$ modulator SC blocks and to the sampled-data circuits in general.

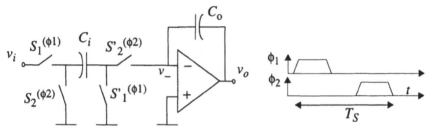

Figure 4.1: SC integrator and clock phases

The knowledge of the functionality of the basic blocks, either by equations [Bose88a] or through tabulated data [Brau90], is essential for the behavior simulation of any system. In the first case, such equations can be simple difference equations that properly represent the output of the system under ideal conditions; for instance, for the integrator of Fig. 4.1

$$v_o(nT_S) = \frac{C_i}{C_o} v_i(nT_S - T_S) + v_o(nT_S - T_S) \tag{4.1}$$

where T_S is the sampling period; or, as will be seen afterward, complex expressions iteratively handled, that include a large amount of non-idealities in order to fit the actual circuit behavior.

An alternative to the use of more or less complex equations is determining the functionality of the basic blocks based on tabulated data from electrical simulation or experimental measurements that are interpolated [Brau90]. Since such tables compile the real circuit behavior, this procedure permits simulation with greater accuracy. However, it represents a loss of generality because tables should be remade after slight changes in the cells that form the basic blocks. Therefore, it is a useful strategy for bottom-up validation, but not very efficient for synthesis.

Once the system is decomposed in basic blocks and their behavioral law is obtained, the simulation process consists of going through the block chain, updating their outputs once each clock cycle.

4.3   BASIC BLOCK BEHAVIORAL MODELS

The fundamental blocks of ΣΔ modulators are integrators, single-bit or multi-bit quantizers and D/A converters, and digital blocks. Through the adequate connection of these few blocks a large number of architectures can be obtained whose behavior can be affected by errors. Let us center on the circuitry with analog part, supposing that the non-ideal behavior of the digital circuitry does not degrade the performance of the modulator.

In addition to these fundamental basic blocks, there are other auxiliary blocks like adders, multipliers, delays, etc., with very simple behavioral models, especially useful for increasing the abstraction of the simulation: for example the system of Fig. 4.2, where the block D represents a delay, maintains the functionality of the integrator of Fig. 4.1 given in (4.1).

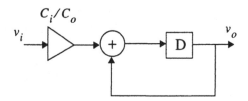

Figure 4.2: Integrator block diagram

4.3.1 Integrator

The integrator can be considered as the fundamental block because, as stated in Chapter 3, due to its position its non-idealities largely affect the operation of the $\Sigma\Delta$ modulators [Cand92]. In particular, using SC techniques, among the causes that worsen the operation of the modulators we find: the incomplete charge transfer, the non-linearity, the thermal noise, etc. [Dias92b][Feel91][Gobe83][Yuka87]; all these are associated with non-idealities of the SC integrator basic cells: amplifier, capacitors and analog switches.

Most such non-idealities were studied in the previous chapter, in order to obtain closed expressions that collect their impact in the modulator operation. Here we take advantage of this knowledge to establish computational models that can be used for behavioral simulation.

4.3.1.1 Sampling phase

Fig. 4.3 shows simulation results obtained using HSPICE [Meta88] corresponding to the extracted layout of an integrator like that of Fig. 4.1 in a fully-differential version[1]. The clock phases and the input are those shown in the figure and the gain of the integrator is $C_i/C_o = 1/4$. As stated previously, the operation of the integrator is clearly divided into two stages, corresponding to each clock phase. During the first one (S_1 and S_1' ON), the input voltage is sampled by the input capacitor C_i, while the output does not change. The charge stored in C_i is transferred, through the switches S_2 and S_2', to the integration capacitor during the phase ϕ_2, so that, at the end of this phase, the output voltage is given by (4.1). The error mechanisms that

---

1. The schematic of this integrator is presented in Section 5.8.2.

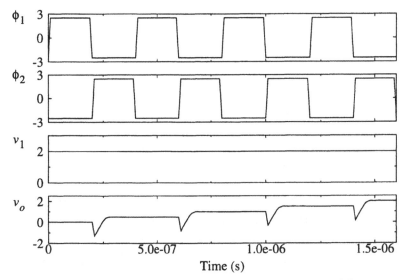

Figure 4.3: Time-domain evolution of the signals in an SC integrator

modify this ideal behavior affect both phases, but it is interesting to isolate their effects. In order to preserve the order of operations, we will begin with those in the sampling phase.

The principal error mechanisms present during the sampling phase are the following:

a) Incomplete settling of the voltage in the input capacitor. A non-zero ON resistance of the analog switches can lead to an RC constant which is excessive in comparison to the duration of the sampling phase. At the end of this phase, the voltage stored in C_i is given by

$$v_i' = v_i\left[1 - \exp\left(-\frac{T_S}{2RC}\right)\right] \tag{4.2}$$

b) Non-linear dependency between the stored charge and the sampled voltage due to capacitor non-linearity.

$$C_i(v_i) = C_i^o(1 + \alpha v_i + \beta v_i^2 + ...) \tag{4.3}$$

c) Thermal noise generated by the ON resistance of the switches. As shown in the previous chapter, the power of this noise is independent of the value of the resistance because the latter acts as a filtering element of its own noise; resulting in the well-known expression kT/C.

These last three non-idealities will be considered in detail during this chapter[†1].

4.3.1.2 Transient response

As shown in the bottom graph of Fig. 4.3, the charge transfer between the capacitors C_i and C_o, ϕ_2 high, is not immediate. On the contrary, if the output resistance of the amplifier is large, the integrator output voltage changes jerkily in the opposite direction to that of the final increase [Sans87], goes through a slew-rate phase, and finally evolves toward its final value according to the dynamics of the amplifier and the involved capacitances. An example of such evolution was shown in Fig. 3.9.

In Section 3.3.1 a single-pole slew-rate model for the integrator transient response was established. The equations obtained there are useful to establish a behavioral model for the transient response; Table 4.1 shows the value of the voltage transferred from the output of the integrator according to the expressions of Section 3.3.1.

Table 4.1: Single-pole slew-rate model for the integrator transient response

if $\|V_i\| \leq \dfrac{I_o}{g_m\varsigma}$ There is no slew-rate		$v_{n-1} + V_i \dfrac{C_i}{C_o}\left[1 - \dfrac{C_i + C_p}{C_i}\varsigma \cdot \exp\left(-\dfrac{g_m}{C_{eq}}\dfrac{T_S}{2}\right)\right]$
if $\|V_i\| > \dfrac{I_o}{g_m\varsigma}$ slew-rate	$\dfrac{T_S}{2} \geq t_0$	$v_{n-1} + V_i\dfrac{C_i}{C_o} - \dfrac{C_i + C_p I_o \operatorname{sgn}(V_i)}{C_o \quad g_m}\exp\left[-\dfrac{g_m}{C_{eq}}\left(\dfrac{T_S}{2} - t_0\right)\right]$
	$\dfrac{T_S}{2} < t_0$	$v_{n-1} + V_i\dfrac{C_i}{C_o} + \left(1 + \dfrac{C_i + C_p}{C_o}\right)\left[\dfrac{I_o \operatorname{sgn}(V_i) T_S}{2 C_{eq}} - \varsigma V_i\right]$

In those expressions

$$C_{eq} = C_i + C_p + C_l\left(1 + \frac{C_i + C_p}{C_o}\right); \quad t_0 = -\frac{C_{eq}}{g_m} + \frac{C_i}{I_o}|V_i|\left(1 + \frac{C_l}{C_o}\right); \quad \varsigma = \frac{C_i}{C_{eq}}\left(1 + \frac{C_l}{C_o}\right)$$

where V_i and v_{n-1} are the voltages stored in the sampling capacitor and inte-

---

1. The effect of the feedthrough is not considered. As shown in [Lee85], by properly timing the phases that control the integrator, it is possible to cancel the signal-dependent charge injection. On the other hand, the signal-independent feedthrough, which translates to an offset, is attenuated using fully-differential architectures.

gration capacitor, respectively, in the beginning of the integration phase.

The existence of a single pole is, of course, an approximation. However, the results obtained with said model fit very well with those obtained through electrical simulation for amplifiers with a large phase margin (60degree or larger), when the second pole is far enough from the first one. For example, Fig. 4.4 shows the output voltage of an integrator like that of Fig. 4.1 during the integration phase. The curve labeled as HSPICE corresponds to the electrical simulation of an integrator extracted layout with the parameters indicated[1]. Note that the curve obtained with the single-pole slew-rate model presents, in the steady state value, very good agreement with the simulated one (the error is smaller than 0.1%).

Nevertheless, this model is inefficient when amplifiers with small phase margin are considered because the influence of higher-frequency poles and zeroes is manifest. On the other hand, it is not always possible, or at least desirable, to design amplifiers with large phase margins, especially for high-frequency applications in which the dynamic properties of the transistors should be fully exploited. For this reason, we will devote the rest of this section to obtain a two-pole model for the transient response of the integrator, which will allow us to simulate integrators with low-phase margin opamps.

Several examples of studies focused on the integrator transient response considering two-pole models are found in literature. However, these studies

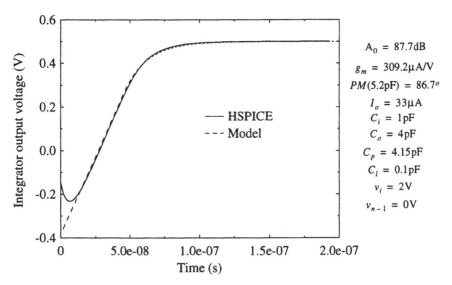

Figure 4.4: Integration-phase output voltage

1. The fully-differential schematic of this amplifier can be found Section 5.8.2.

correspond to integrators with two-stage amplifiers. Recently, the single-stage amplifiers (folded-cascode type, Fig. 4.5) [Ribn84], due to their good performance-consumption relationship, have replaced those with two gain stages in several analog signal processing applications. We will consider, therefore, this type of amplifier. Its dynamic behavior is similar to that of the conceptual model of Fig. 4.6(b). This model has two stages, though only one is a gain stage. Both stages can be considered as non-coupled because the gate voltage of the transistors M_A and M_B is fixed. Furthermore, the model incorporates two voltage-controlled non-linear current sources to contemplate the current saturation of both stages.

Considering both stages in linear operation, the Laplace-domain transfer function of the circuit of Fig. 4.6(b) is given by

$$\frac{V_o(s)}{V_a(s)} = -\frac{(g_{m1}/C_1)(g_{m2}/C_2)}{(s+g_1/C_1)(s+g_2/C_2)} \tag{4.4}$$

We will assume that, like in a typical design, out of the two poles in (4.4), the one corresponding to the output stage is the dominant pole. With respect to the model of the complete integrator, the capacitor C_2 includes the parasitic at the output of the amplifier and a possible load at the integrator output.

To simplify our analysis we will suppose that i_1 stays always in its linear region so that the slew-rate is caused by the output stage. Thus, for a voltage level, V_i, stored in C_i, in general, the voltages v_a, v_1 and v_o evolve linearly until the saturation of i_2 takes place. After that, this source operates as an independent current source with a value equal to its saturation level. Such a situation continues until the absolute value of v_1 decreases, so that i_2 enters again in its linear region. In what follows, these three situations will be analyzed separately.

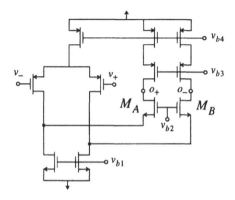

Figure 4.5: Fully-differential folded-cascode amplfier

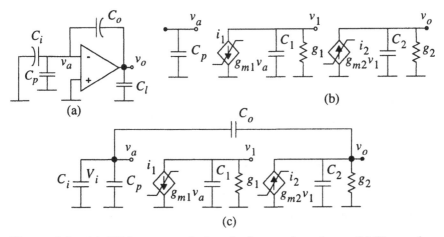

Figure 4.6: (a) SC integrator during the integration phase. (b) Two-pole model for the amplifer. (c) Complete integrator model.

1. First part of the transient response: both stages in linear region

Let V_i be the voltage stored in C_i during the sampling phase. We will suppose for simplicity that the rest of the capacitors are uncharged. In fact, an arbitrary initial condition in C_o and C_2 does not affect the validity of the model. Resolving the equations that describe the circuit of Fig. 4.6(c) in the Laplace domain, yields the following expressions:

$$V_o(s) = \frac{C_i C_o}{\gamma_C} \cdot \frac{s^2 + \dfrac{g_1}{C_1}s - \dfrac{g_{m1}g_{m2}}{C_1 C_o}}{s^2 + 2\alpha s + \alpha^2 + \beta^2} \cdot \frac{V_i}{s} \tag{4.5}$$

$$V_1(s) = \frac{g_{m1} C_i (C_o + C_2)}{C_1 \gamma_C} \cdot \frac{s + \dfrac{g_{m1}}{C_o + C_2}}{s^2 + 2\alpha s + \alpha^2 + \beta^2} \cdot \frac{V_i}{s}$$

$$V_a(s) = \frac{C_i (C_o + C_2)}{\gamma_C} \cdot \frac{\left(s + \dfrac{g_2}{C_o + C_2}\right)\left(s + \dfrac{g_1}{C_1}\right)}{s^2 + 2\alpha s + \alpha^2 + \beta^2} \cdot \frac{V_i}{s}$$

where

$$\gamma_C = C_i C_2 + C_i C_o + C_2 C_o + C_2 C_p + C_o C_p$$

$$C_a = C_i + C_p + C_o$$

$$\alpha = \frac{g_2 C_a}{2\gamma_C} + \frac{g_1}{2C_1} \tag{4.6}$$

$$\beta = \sqrt{\frac{g_{m1}g_{m2}C_o}{C_1\gamma_C} + \frac{g_1 g_2 C_a}{2C_1\gamma_C} - \frac{g_1^2}{4C_1^2} - \frac{g_2^2 C_a^2}{4\gamma_C^2}}$$

We will consider the case in which the poles of the complete system are imaginary; that is, the member under the square root in (4.6) is positive. Taking an inverse Laplace transform in (4.5) gives the following time-domain expressions for the three voltages:

$$v_o(t) = A_{o1} + B_{o1}\exp(-\alpha t)\cos\beta t + C_{o1}\exp(-\alpha t)\sin\beta t$$

$$v_1(t) = A_{11} + B_{11}\exp(-\alpha t)\cos\beta t + C_{11}\exp(-\alpha t)\sin\beta t \tag{4.7}$$

$$v_a(t) = A_{a1} + B_{a1}\exp(-\alpha t)\cos\beta t + C_{a1}\exp(-\alpha t)\sin\beta t$$

Due to their complexity, it is difficult to expand these expressions. Instead, Table 4.2 shows the value of the coefficients in (4.7). Each coefficient is obtained multiplying the common term by the expression in the corresponding cell. These relationships will be valid while the condition

$$g_{m2}|v_1(t)| \le I_o \tag{4.8}$$

is fulfilled, where I_o is the saturation current of the second stage. If this inequality is maintained, the settling of the integrator output voltage will be linear and the solutions obtained, (3.7), valid for the whole integration phase.

We will suppose a more realistic generic case in which the second stage saturates. The following step consists of calculating the instant in which the equality is reached in (3.8); that is, the time, t_1, in which the slew-rate period begins. To do so, the following equation should be solved

$$v_1(t_1) = A_{11} + B_{11}\exp(-\alpha t_1)\cos\beta t_1 + C_{11}\exp(-\alpha t_1)\sin\beta t_1 = \frac{I_o}{g_{m2}} \tag{4.9}$$

Unfortunately such an equation contains transcendent functions and cannot be solved by algebraic methods. Therefore, a numerical method has to be used to obtain the value of t_1. We suppose that the values of the voltages in t_1, that we will call V_{a1}, V_{11} y V_{o1}, are known.

Table 4.2: Coefficients in (4.7)

Coefficient	Common term	
A_{o1}	$\dfrac{C_iC_o}{\gamma_C}V_i$	$(g_{m1}g_{m2})/[(\alpha^2+\beta^2)C_oC_1]$
B_{o1}		$1-(g_{m1}g_{m2})/[(\alpha^2+\beta^2)C_oC_1]$
C_{o1}		$\dfrac{\alpha}{\beta}\left(\dfrac{g_{m1}g_{m2}}{(\alpha^2+\beta^2)C_oC_1}-1\right)+\dfrac{g_1}{\beta C_1}$
A_{11}	$\dfrac{g_{m1}C_iV_i}{\gamma_C C_1}$	$g_2/(\alpha^2+\beta^2)$
B_{11}		$-g_2/(\alpha^2+\beta^2)$
C_{11}		$\dfrac{C_o+C_2}{\beta}-\dfrac{\alpha}{\beta}\dfrac{g_2}{(\alpha^2+\beta^2)}$
A_{a1}	$\dfrac{C_i(C_o+C_2)V_i}{\gamma_C}$	$g_1g_2/[(\alpha^2+\beta^2)C_1(C_o+C_2)]$
B_{a1}		$1-g_1g_2/[(\alpha^2+\beta^2)C_1(C_o+C_2)]$
C_{a1}		$\dfrac{1}{\beta}\left(\dfrac{g_1}{C_1}+\dfrac{g_2}{C_o+C_2}\right)-\dfrac{\alpha}{\beta}\left(1+\dfrac{g_1g_2}{(\alpha^2+\beta^2)C_1(C_o+C_2)}\right)$

2. *Second part of the transient response: second stage saturated*

The model for the circuit of Fig. 4.6(a) valid during this part of the transient response is shown in Fig. 4.7, where the initial conditions in the capacitors have been introduced as independent voltage sources. Solving this circuit in the Laplace domain, yields

$$V_o(s) = V_{o1}\cdot\frac{s+I_oC_a/(V_{o1}\gamma_C)}{s(s+g_2C_a/\gamma_C)}$$

$$V_1(s) = \frac{I_o}{g_{m2}}\cdot\frac{s^2+\left(\dfrac{g_2C_a}{\gamma_C}-\dfrac{g_{m1}g_{m2}V_{a1}}{C_1I_o}\right)s-\dfrac{g_{m1}g_{m2}}{C_1\gamma_CI_o}(C_oI_o+g_2C_aV_{a1}-g_2C_oV_{o1})}{s(s+g_1/C_1)(s+g_2C_a/\gamma_C)}$$

$$V_a(s) = V_{a1}\cdot\frac{s+(C_oI_o+g_2C_aV_a-g_2C_oV_o)/(\gamma_CV_{a1})}{s(s+g_2C_a/\gamma_C)}$$

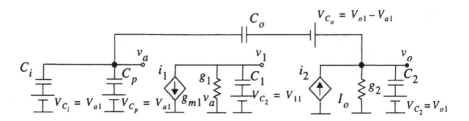

Figure 4.7: Slew-rate phase model

where V_{11}, which represents the initial condition in C_1, has been substituted by its value I_o/g_{m2}. Taking an inverse Laplace transform give us the following time-domain expressions for the three voltages:

$$v_o(t) = A_{o2} + B_{o2}\exp[-p_2(t-t_1)]$$
$$v_1(t) = A_{12} + B_{12}\exp[-p_1(t-t_1)] + C_{12}\exp[-p_2(t-t_1)] \qquad (4.11)$$
$$v_a(t) = A_{a2} + B_{a2}\exp[-p_2(t-t_1)]$$

where

$$p_1 = \frac{g_1}{C_1} \qquad p_2 = \frac{g_2 C_a}{\gamma_C} \qquad (4.12)$$

and the coefficients are given in the Table 4.3. The slew-rate will continue until the equality in (4.8) is reached. To find this time, t_2, the following equation should be solved

$$v_1(t_2) = A_{12} + B_{12}\exp[-p_1(t_2-t_1)] + C_{12}\exp[-p_2(t_2-t_1)] = \frac{I_o}{g_{m2}} \qquad (4.13)$$

once again by using numerical methods. At the moment, we will suppose that t_2 is known.

3. *Third part of the transient response: second stage in linear region again*

For $t > t_2$ the current source in the output stage enters in its linear operation region, with which the model of Fig. 4.6(c) is valid once more. Only the initial conditions will have changed. Solving this circuit with

$$V_{C_o} = V_{o2} - V_{a2} \qquad V_{C_i} = V_{a2} \qquad V_{C_p} = V_{a_2}$$
$$V_{C_1} = V_{12} = \frac{I_o}{g_{m2}} \qquad V_{C_2} = V_{o2} \qquad \text{where } V_{x2} \equiv v_x(t_2) \quad (4.14)$$

Table 4.3: Coefficients in (4.11)

Coefficient	Common term	
A_{o2}	1	I_o/g_2
B_{o2}		$(g_2V_{o1}-I_o)/g_2$
A_{12}		$\dfrac{(g_2C_oV_{o1}-g_2C_aV_a-C_oI_o)}{g_1g_2C_a}$
B_{12}	g_{m1}	$\dfrac{\left[C_o+\dfrac{p_1C_1(g_2C_a-\gamma_Cp_1)}{g_{m1}g_{m2}}\right]I_o+(g_2C_a-\gamma_Cp_1)V_{a1}-g_2C_oV_{o1}}{\gamma_Cp_1(p_2-p_1)C_1}$
C_{12}		$\dfrac{\left[C_o+\dfrac{p_2C_1(g_2C_a-\gamma_Cp_2)}{g_{m1}g_{m2}}\right]I_o+(g_2C_a-\gamma_Cp_2)V_{a1}-g_2C_oV_{o1}}{\gamma_Cp_2(p_1-p_2)C_1}$
A_{a2}	1	$\dfrac{C_oI_o+g_2C_aV_{a1}-g_2C_oV_{o1}}{g_2C_a}$
B_{a2}		$\dfrac{C_o}{g_2C_a}(g_2V_{o1}-I_o)$

as new initial conditions, the following expression is obtained for $V_o(s)$:

$$V_o(s) = \frac{V_{o2}}{s} \cdot \frac{s^2+\left(\dfrac{C_aI_o}{\gamma_CV_{o2}}+\dfrac{g_1}{C_1}\right)s+\dfrac{g_{m1}g_{m2}}{\gamma_CC_1V_{o2}}(C_oV_{o2}-C_aV_{a2})}{s^2+2\alpha s+\alpha^2+\beta^2} \tag{4.15}$$

and hence

$$v_o(t) = A_{o3}+B_{o3}\exp(-\alpha t)\cos\beta t+C_{o3}\exp(-\alpha t)\sin\beta t \tag{4.16}$$

Table 4.4: Coefficients in (4.16)

Coefficient	Common term	
A_{o3}		$g_{m1}g_{m2}\dfrac{(C_oV_{o2}-C_aV_a)}{[(\alpha^2+\beta^2)\gamma_CC_1V_{o2}]}$
B_{o3}	V_{o2}	$1-g_{m1}g_{m2}\dfrac{(C_oV_{o2}-C_aV_a)}{[(\alpha^2+\beta^2)\gamma_CC_1V_{o2}]}$
C_{o3}		$\dfrac{\alpha}{\beta}\left[\dfrac{g_{m1}g_{m2}(C_aV_a-C_oV_o)}{(\alpha^2+\beta^2)\gamma_CC_1V_{o2}}-1\right]+\dfrac{g_1}{\beta C_1}+\dfrac{C_aI_o}{\beta\gamma_CV_{o2}}$

Table 4.4 contains the coefficients in (4.16). The rest of the voltages are unimportant because we assume that the current status does not change until the end of the integration phase.

4. *Computational algorithm*

As mentioned previously, to calculate t_1 and t_2 it is necessary to use numerical methods. Because of that, this section is devoted to obtaining a calculation algorithm, based on the previous equations, that can be easily adapted to any programming language. Fig. 4.8 shows a flow graph of said algorithm aimed at finding the value of the integrator output voltage at the end of the sampling phase. In the *Calculate* blocks the number of order of the equation to be used is specified. The shading blocks indicate iterative calculations. A possibility for the calculation of t_1 and t_2 is to use the Newton-Raphson method. As an example, Fig. 4.9 shows the corresponding flow graph to calculate t_2. Generally, convergence is detected after very few iterations.

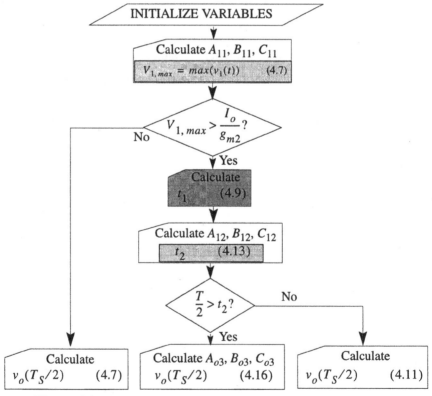

Figure 4.8: Flow graph to estimate the integrator output voltage

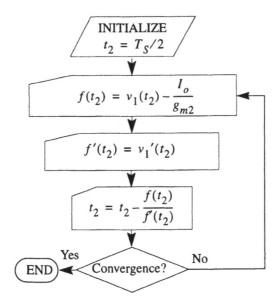

Figure 4.9: Flow graph for the calculation of t_2

To prove the effectiveness of the calculations, a real amplifier design with folded-cascode fully-differential architecture like that of Fig. 4.5[1] has been used. Such an amplifier was intentionally designed with a small phase margin (approximately 45degree) in order to reduce the power consumption. Fig. 4.10 shows the integrator output voltage during the integration phase obtained through electrical simulation (HSPICE) and with the procedure in Fig. 4.8 using the parameters indicated. Both approaches go together very well.

A more global vision is given in Fig. 4.11, which shows the difference between the output voltage at the end of the integration phase and the corresponding ideal final value, for several levels of the input. Three curves corresponding to three durations of the integration phase 11, 15 and 18ns are represented.

Once more the coincidence of the electrical simulation and the behavior obtained with the model is notable. On the other hand, Fig. 4.11 shows that in some cases amplifiers with *PM* < 60degree can be used with no excessive degradation of the modulator performance. As will be seen in Section 4.5.2 and in Chapter 7, this consideration has useful results in the design of high-frequency, low-power consumption ΣΔ modulators.

---

1. Details on this amplifier can be found in Section 7.4.6.

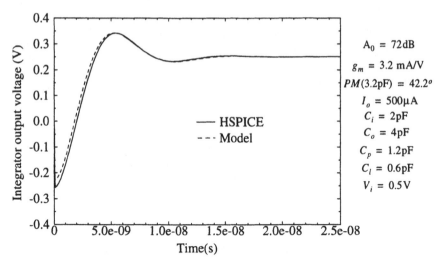

Figure 4.10: Integrator output voltage during the integration phase

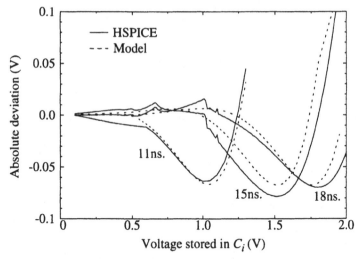

Figure 4.11: Deviation of the output voltage in respect to its ideal value for three durations of the integration phase, as a function of the input level

4.3.1.3 Thermal noise

Because the time-domain behavioral simulation is event-driven, only those phenomena that present changes with a frequency smaller or equal to that of the clock can be modeled. Furthermore, the modeling of continuous nature mechanisms is possible only when their influence is circumscribed to each individual clock cycle, or in other words, such influence can be synchronously partitioned. This is the case, for example, of the integrator settling covered in the previous section.

This does not occur with thermal noise: first, it is a phenomenon of continuous nature that varies randomly and second, its cut-off frequency is generally well above the sampling frequency. The first problem is partially soluble because, though computers base their operation on deterministic algorithms, it is possible to build routines that generate random numbers with very low self-correlation. So, the principal difficulty is the timing: the reduction of the internal time-step of a computation algorithm in order to adapt it to the change speed of the signals it handles, is a feature of numerical simulators. Thus, the calculation speed – the main advantage of the behavioral simulation, is definitively lost.

In the particular case of the oversampling converters, there exists an approximation to the problem more related to the behavioral simulation. Since the clock frequency is several times larger than that of the signals being processed, the white noise modulation leads to an input-equivalent noise that, at least in the frequencies of interest, presents a constant spectral density. The modulation of this noise, patent by the presence of a *Sinc* function in expressions like (3.69), is effective only for frequencies near the sampling frequency. Taking advantage of this fact, the modulator input-equivalent thermal noise can be pre-calculated and injected at the clock rate using a random number generator. Thus, a power spectral density is obtained that equals that of the modulated thermal noise in the low-frequency region.

In Chapter 3 expressions were obtained to estimate the thermal noise power spectral density of an SC integrator like that of Fig. 4.12. The generic equation, which takes into account not only the noise generated in the ON resistances of the switches but also that of the amplifier, is

$$
S_{eq, C_{11}}(f) \cong \left(\frac{\tau_{SH}}{T_S}\right)^2 \left[\left(1 + \frac{C_{12}}{C_{11}}\right)\frac{kT}{f_S C_{11}} + \left(1 + \frac{C_{12}^2}{C_{11}^2}\right)\frac{2kT}{3 f_S C_i}\right]
$$
$$
+ \left(\frac{\tau_T}{T_S}\right)^2 \left(1 + \frac{C_{12}^2}{C_{11}^2}\right)\frac{2kT g_m R_{on}}{f_S C_i}
$$

(4.17)

where $C_i = C_p + C_{11} + C_{12}$, with C_p equal to the integrator input parasitics and $\tau_T = T_S - \tau_{SH}$ equal to duration of the phase ϕ_1.

Thus, in each period, the extra voltage stored in the sampling capacitor that adequately represents the equivalent thermal noise of the modulator is

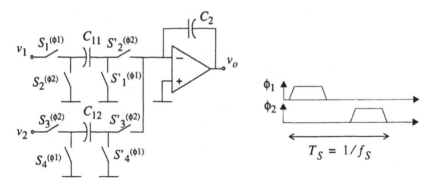

Figure 4.12: Two-branch SC integrator and clock phases

given by

$$v_{th} = \text{rnd}(\sqrt{12 f_S S_{eq, C_{11}}}) \tag{4.18}$$

where $\text{rnd}(x)$ represents a random number in the range $(-x/2, x/2)$ and it has been applied that the power of a signal with constant probability density in such a range is $x^2/12$, [Benn48].

4.3.1.4 Non-linearity

Several error mechanisms associated with the integrator gain non-linearity were studied in Chapter 3. One of them is due to the slew-rate in the settling of the output voltage and has been already covered in the behavioral simulation context. We will devote this section to the remaining two causes of distortion: non-linear capacitance and non-linear open-loop gain of the amplifier.

1. *Capacitor non-linearity*

Ideally, the finite-difference equation that represents the behavior of an SC integrator (Fig. 4.12) is

$$v_{o, n} C_2 = v_{1, n-1} C_{11} - v_{2, n} C_{12} + v_{o, n-1} C_2 \tag{4.19}$$

If we consider that the capacitors are non-linear; that is, that their capacitance varies with the stored voltage,

$$C(v) = C^o (1 + \alpha v + \beta v^2 + ...) \tag{4.20}$$

according to Section 3.5.1, equation (4.19) is rewritten as

$$(4.21)$$

$$v_n = \left[v_{1,n-1} C_{11}^o \left(1 + \frac{\alpha}{2} v_{1,n-1} + \frac{\beta}{3} v_{1,n-1}^2 \right) - v_{2,n} C_{12}^o \left(1 + \frac{\alpha}{2} v_{2,n} + \frac{\beta}{3} v_{2,n}^2 \right) \right.$$
$$\left. + v_{n-1} C_2^o \left(1 + \frac{\alpha}{2} v_{n-1} + \frac{\beta}{3} v_{n-1}^2 \right) \right] \Big/ C_2^o \left(1 + \frac{\alpha}{2} v_n + \frac{\beta}{3} v_n^2 \right)$$

where v_o has been substituted by v for simplicity. Note that the numerator is known once the value of $v_{1,n-1}$, $v_{2,n}$, and v_{n-1} is determined, while the denominator depends on v_n. Because of that, this equation must be solved numerically: however, for weak non-linearity, as observed in practice, where $\alpha \ll 1$ y $\beta \ll 1$, a simple fast-convergence iterative procedure permits the calculation of the final value of v_n.

2. *Opamp open-loop gain non-linearity*

A similar procedure can be used to simulate the dependency of the amplifier open-loop gain with its output. Considering a finite gain model for the amplifier of Fig. 4.12, the operation of the integrator can be approximated by

$$v_n \cong \frac{A_V v_{i,n-1}}{A_V + 1 + g_1(1+k_1)} + \frac{(A_V+1)v_{n-1}}{A_V + 1 + g_1(1+k_1)} \qquad g_1 = \frac{C_{11}}{C_2} \qquad k_1 = \frac{C_{12}}{C_{11}} \qquad (4.22)$$

where $v_i = g_1(v_1 \pm k_1 v_2)$. Supposing that the gain of the amplifier varies with the output voltage such as

$$A_V = A_0(1 + \gamma_1 v_n + \gamma_2 v_n^2 + \ldots) \qquad (4.23)$$

iterating the expressions (4.22) and (4.23) leads quickly to the final value of the v_n after a few steps.

3. *Behavioral model for both non-linearities*

The effect of the capacitor and opamp open-loop gain non-linearity are included simultaneously in the flow graph of Fig. 4.13. After initializing the input voltages, the integrator internal state and the amplifier gain to their nominal values, the value of v_{n-1}, $v_{1,n-1}$ and $v_{2,n}$ is modified according to the current value of A_V. Then, the value of v_n is calculated using (4.21), supposing in addition that the denominator of this expression equals the unity. Subsequently v_n is updated for the actual values of α and β and, finally, the new value of the open-loop gain is calculated with (4.23). The convergence of this procedure is detected after two or three iterations which does not result in excessively costly CPU time.

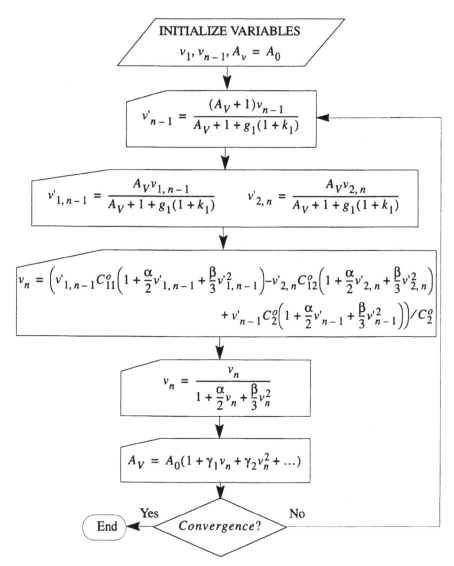

Figure 4.13: Calculation of the integrator output voltage in presence of non-linearity

4.3.1.5 Jitter noise

In Section 3.5.3, the impact of the dispersion of the clock period in SC $\Sigma\Delta$ modulators operation was analyzed. This phenomenon, known as Jitter, leads to an important degradation of the quantization noise-shaping function. The larger the frequency of the input signal, the larger is such degradation. Though this error mechanism is due to the integrator, in terms of the behavioral simulation it is more useful to consider it solidary to the sinusoidal signal generator; that is, a pure sinusoidal signal as input to a modulator with Jitter is equivalent to a sinusoidal signal, previously contaminated by said mechanism, serving as input to an ideal modulator. Note that such equivalence implies that, as for other error mechanisms, only the contribution of the first integrator is taken into account.

Mathematically, the modulator input voltage provided by the sinusoidal generator with Jitter can be expressed as follows:

$$v_i(nT_S) = A\sin[2\pi(nT_S + \Delta t)] \tag{4.24}$$

where Δt is adjusted to a Gaussian distribution with zero average and standard deviation σ_t:

$$\Delta t = gauss(0, \sigma_t) \tag{4.25}$$

4.3.1.6 Complete integrator model

Fig. 4.14 shows a flow graph of the operations in the complete integrator model. In this, all the studied processes take place together with others of simpler nature, as the limitation of the output range. The flow graph has two branches corresponding to the two clock phases:

- During the sampling phase, the final value of the voltages in the input capacitors are stored taking into account the value of these and the ON resistance of the switches.

- During the integration phase, the input-equivalent thermal noise of the integrator is calculated and added to the sampled voltage. Next, an iterative procedure is started to calculate the integrator output voltage, including the effects of the finite and non-linear opamp, the non-linear capacitors, transient response and the limitation of the output range. The convergence of this process is reached after two or three iterations.

4.3.2 Quantizers

Most popular ΣΔ modulator architectures incorporate as a quantizer ele-

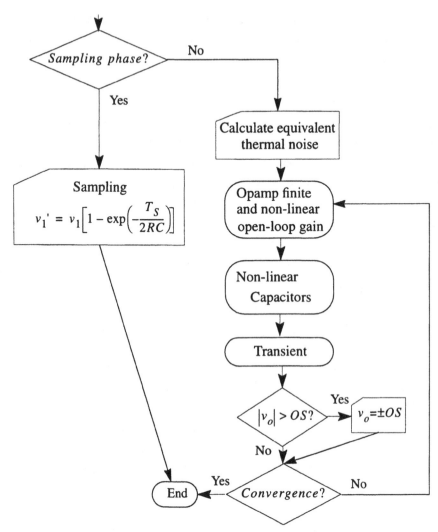

Figure 4.14: Complete integrator model

ment a simple comparator or single-bit quantizer. The loss of resolution caused by such gross quantization is widely compensated because a single-bit D/A converter can be used in the feedback loop, which, in addition to its simplicity, is a perfectly linear block and does not introduce any non-linearity error. Nevertheless, novel techniques, such as dynamic calibration [Chen95][Bair96][Nys96] or the selection of proper architectures [Bran91b] have corrected, in part, the distortion problems caused by the non-linearity of the multi-bit D/A converter, so that these can be used in certain applications.

These blocks are studied below in terms of the behavioral simulation.

4.3.2.1 Comparator

The impact of the comparator non-idealities in the operation of the ΣΔ modulators is much lower than those of the integrator. This is due to the position that the comparator occupies in the modulator loop. For example, a possible input offset in the comparator is attenuated by the DC gain(s) of the integrator(s) that precede it in the loop. Because of that, especially when the order (or number of integrators in the loop) increases, the modulators are practically insensitive to such an error. The same reasoning can be applied to other non-idealities, such as the hysteresis of the comparator whose error receives the same treatment as the quantization noise: in practice, comparator resolutions so coarse as 10% of the full-scale range do not degrade significantly the modulator performance. On the other hand, in this case, there are no problems associated with the D/A conversion non-linearity as stated previously.

The comparator hysteresis leads to a loss of resolution due to the fact that, for signals next to the comparison threshold, there exists a resistance to change the output state even when the input level may have surpassed such a threshold. This phenomenon, that is appreciated in both state changes generating a hysteresis cycle (see Fig. 4.15), is because the comparator has a memory of the previous state, an overdrive being necessary to make it commute to the correct state. In addition to this type of hysteresis, which we could call deterministic, there is another of random nature. It is a typical error of the latched comparators with a re-set phase during which the memory of the previous state must be fully eliminated. Due to deficiencies of the real devices, this process lets in certain residues of the previous state, leading to an uncertainty zone (see Fig. 4.16). In this zone the output of the comparator is not

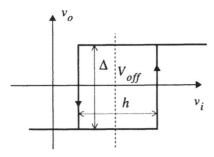

Figure 4.15: Transfer curve of a comparator with hysteresis

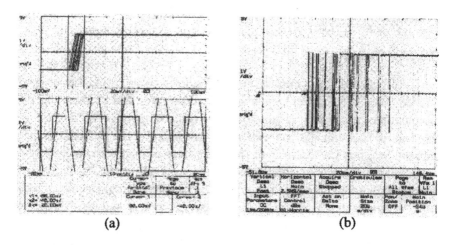

Figure 4.16: Measurement of the hysteresis of a latched comparator: (a) Transfer curve and time-domain evolution of input and output (b) transition detail.

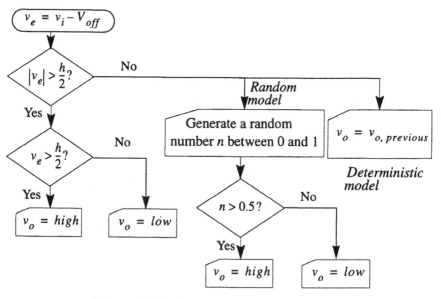

Figure 4.17: Comparator model flow graph

determined only by its input, but, in addition, by the previous state and the transitions of the signals re-setting the latch.

To contemplate the latter, the comparator model of Fig. 4.17 presents three well-differentiated operation zones. In the first of them, in response to an input v_i, the output value (low or high) is a function of the sign of v_i, provided that $|v_i - V_{off}| > h/2$ is fulfilled, where V_{off} and h represent the offset and the hysteresis of the comparator, respectively. Otherwise, the output is randomly determined in the dynamic hysteresis model and simply does not change in the deterministic model. With respect to the analytical model, Bosser and Wooley [Bose88b] show that, in an Lth-order modulator, the in band power of the error due to the comparator hysteresis can be approximated by $P_h = 4h^2\pi^{2L}/[(2L + 1)M^{(2L + 1)}]$, which fits very well with the result obtained through behavioral simulation with the deterministic model, as well as with the random model.

4.3.2.2 Multi-bit quantizer and D/A converter

The modulators with more than two internal quantization levels, also called multi-bit modulators, are to a large extent insensitive to the quantizer non-idealities. However, the D/A conversion of the signals in the feedback loop, when two or more bits are handled, can be affected by non-linearity error. In some architectures, the D/A conversion error is directly added to the modulator input and it appears at the output, generally, as distortion. Since such errors are not attenuated, the linearity of the ΣΔ converter is subordinated to that of internal D/A converter, which considerably hinders the design. Nevertheless, the appearance of novel techniques [Sarh93][Chen92] [Bair95][Nys96][Lesl90][Kiae93], that have allowed to palliate to a point this extreme sensibility, has favored the use of multi-bit architectures and makes their treatment interesting in terms of behavioral simulation.

A previous step towards the modeling of multi-bit A/D and D/A converters is the analysis of the possible error mechanisms in their transfer curve. For example, the Fig. 4.18 shows these errors for a 3-bit A/D/A system. Note that the transfer curves of Fig. 4.18(b) are analog-analog; that is, they correspond to the complete A/D/A system.

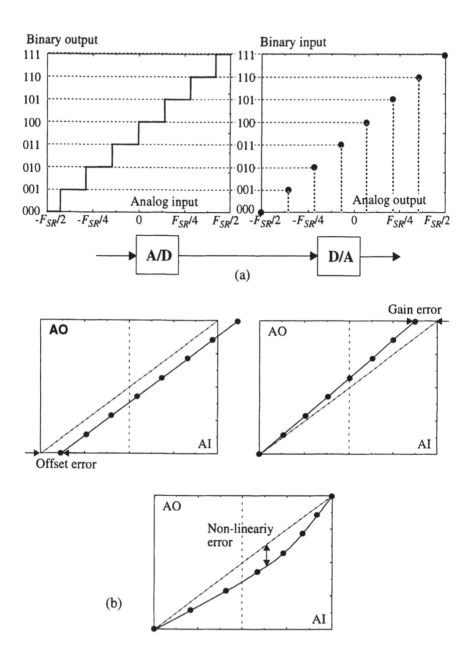

Figure 4.18: (a) 3-bit A/D/A system. (b) Errors in the analog-input ana-
log-output transfer curve.

Among the errors shown: offset, gain and non-linearity; the latter is the one which presents larger problems in a practical case. We will represent the non-linearity of a A/D or D/A converter by its integral non-linearity (*INL*), usually defined as follows [Jede89]:

- For an A/D converter, maximum difference between the analog value which is the actual threshold between two adjacent levels and its ideal value, once the offset and gain errors have been corrected, see Fig. 4.19(a).

- For a D/A converter, maximum difference between the actual output analog value and its ideal value, once the offset and gain errors have been corrected, see Fig. 4.19(b).

Note that, while the gain and offset errors fully specify the gain and the offset of the converter (see Fig. 4.18), there can be many converters whose transfer curves show the same *INL*. To solve this ambiguity we will suppose that the converters present a third-order non-linearity. Thus, the models used are those shown in Fig. 4.20.

For the A/D converter, first an offset *off* is added to the ideal input x_a, and they are multiplied by a factor γ. The result serves as input to a non-linear block with transfer function $\Phi(.)$. Finally, the contaminated input y_a is quantized by an ideal A/D converter. For the D/A converter the model is the dual of the previous: first, the digital input x is translated to the analog plane through an ideal D/A converter. The result x_a goes through the non-linear

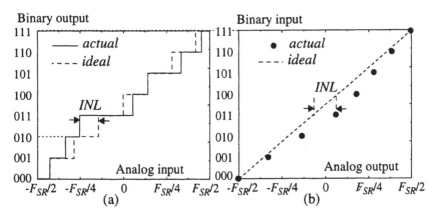

Figure 4.19: Illustrating the *INL* concept for a 3-bit (a) A/D converter and (b) D/A converter.

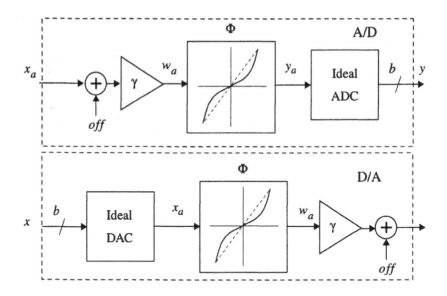

Figure 4.20: Model for the behavioral simulation of A/D and D/A converters

block and finally the gain and offset errors are introduced. The parameters Φ, γ and *off* are calculated for the A/D as follows:

$$y_a = \Phi(w_a) = (1 - \varepsilon_0)w_a + \frac{\varepsilon_0}{A^2}w_a^3 \qquad \varepsilon_0 = \frac{\sqrt{27}}{2^b - 2}INL$$

$$w_a = \gamma(x_a + off) \qquad A = (2^{b-1} - 1)q$$

$$\gamma = 1/(1 + q\varepsilon_g) \qquad off = (l_1\varepsilon_g - \varepsilon_{off})q \qquad l_1 = -F_{SR}/(2G) + q/2$$

$$(4.26)$$

where b is the number of bits and q is the distance between analog values corresponding to adjacent code steps, also called Less Significative Bit (*LSB*); it is obtained dividing the full scale F_{SR} by the number of levels $2^b - 1$. The parameter G denotes the nominal gain of the converter that can be different from the unity. For the D/A converter we have

$$y_a = \gamma(w_a + off) \qquad \gamma = 1/(1 + q\varepsilon_g)$$

$$off = (l_1\varepsilon_g - \varepsilon_{off})q \qquad l_1 = -F_{SR}/(2G)$$

$$w_a = \Phi(x_a) = (1 - \varepsilon_0)x_a + \frac{\varepsilon_0}{A^2}x_a^3 \qquad \varepsilon_0 = \frac{\sqrt{27}}{2^b - 2}INL$$

$$A = (2^{b-1} - 1)q$$

$$(4.27)$$

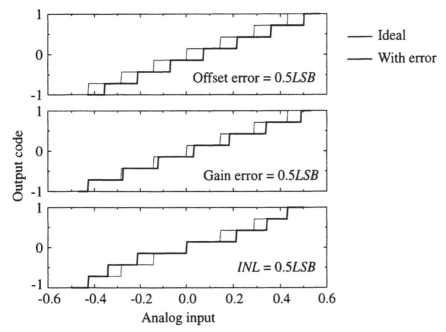

Figure 4.21: Transfer curve of a 3-bit A/D converter obtained with the model including 0.5LSB offset, gain and non-linearity error

In these expressions the units of the gain error ε_g, offset error ε_{off} and of non-linearity error *INL* are *LSB*s. For example, Fig. 4.21 shows the obtained transfer curve including 0.5*LSB* of offset, gain and non-linearity error, in the model of the 3-bit A/D converter.

4.4   *ASIDES*: A TOOL FOR BEHAVIORAL SIMULATION OF SC ΣΔ MODULATORS

4.4.1   Description of the tool

Fig. 4.22 shows a block diagram of the proposed tool. ASIDES allows the behavioral simulation of arbitrary SC ΣΔ modulator architectures. For this, it includes a basic block library that can be considered ideal, with a simple behavioral law, or affected by errors derived from their physical implementation. Table 4.5 compiles the fundamental basic blocks and the non-idealities covered in the simulator. These are included in models associated with each block, with the possibility of defining as many models as necessary for the

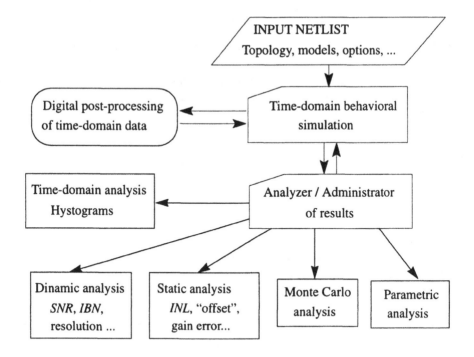

Figure 4.22: Simulator block diagram

same class of basic block.

ASIDES starts from a netlist where the topology and a set of non-idealities to take into account during the simulation are described, and operates in the time-domain using the basic block functional descriptions developed in the previous section. This process generates time-domain series that are digitally processed using MATLAB [Math91] or other signal processing software to perform the following types of analysis:

1. *Time-domain analysis*

Allows visualization of the time-domain waveforms and generates histograms.

2. *Dynamic analysis*

The dynamic characterization of the converters is especially important for medium/high-frequency applications. The dynamic analysis includes the output spectrum; signal - (noise + distortion) ratio (*TSNR*) curves as a function of the input level or frequency, etc.; similar curves for the in band error

Table 4.5: Fundamental blocks and non-idealities included in ASIDES

Basic block		Non-ideality	Consequences
SC Integrator / Opamps	Opamps	Finite and non-linear gain	Quantization noise increase, harmonic distortion
		dynamic limitations	Incomplete settling noise, harmonic distortion
		Output range	Overloading, harmonic distortion
		Thermal noise	White noise
	Switches	ON resistance, thermal noise	Incomplete settling noise, white noise
	Capacitors	non-linearity, mismatching	Quantization noise increase, harmonic distortion
Clock		Jitter	Jitter noise
Comparators		Hysteresis, offset	Quantization noise increase
Quantizers / D-A		non-linearity, gain error, offset error	Quantization noise increase, harmonic distortion

power, effective resolution, etc.

3. *Static analysis*

Important in low-frequency applications, it allows evaluation of the DC transfer curve of the converter and comparison with the ideal one. ASIDES calculates the code step best fitting line of the simulated curve and compares it with the ideal in order to calculate the offset, gain and non-linearity error.

4. *Monte Carlo analysis*

This analysis takes into account fluctuations of the integrators gains and/or of the basic block specifications. The standard deviation of the integrator gain variations can be directly indicated by the user, or calculated using the expression (3.23). This capability is especially interesting when cascade architectures are considered, due to their sensitivity to the mismatching. On the other hand, the terminal specifications of the basic blocks, once designed at the electrical level, can experience changes due to electrical parameter variations, temperature changes, etc. These changes can be included in a Monte Carlo analysis to prove their impact on the modulator performance.

5. *Parametric analysis*

Through this analysis it is possible to perform some of the previous analyses with non-idealities defined as *variable parameters*. The aim is to explore the design space, monitoring individually or jointly the influence of critical design parameters.

4.4.2 Programming features

ASIDES is written in C language, it has about 10,000 code lines and runs under the UNIX operative system. The internal organization of the tool is simple, as illustrated in the flow graph of Fig. 4.23. All the basic blocks are represented through routines that describe their functionality. Said routines are invoked according to the connectivity expressed by the user and update the voltage of a node (output node of the basic block) according to its input node voltage, its internal state and the non-idealities included in the associ-

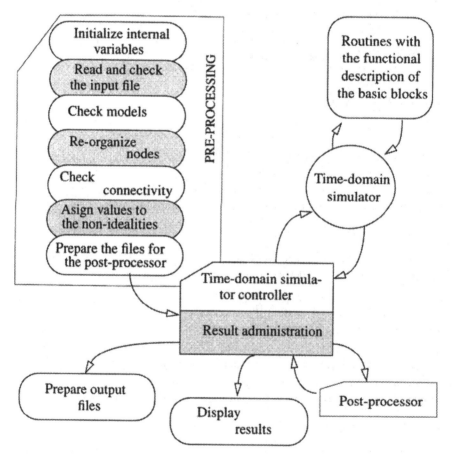

Figure 4.23: ASIDES flow diagram

ated model.

4.5   *ASIDES*: EXAMPLES AND ILLUSTRATIVE RESULTS

The possibility of including arbitrary topologies in ASIDES, together with the large number of analyses that the tool offers, makes it impossible to itemize all its different capabilities. Instead of that, in this section we will center on two examples that we consider representative, either because of their extended use, as a single-loop second-order modulator, or because it is a novel architecture not completely covered by existing simulation tools, as a cascade modulator with multi-bit quantization.

4.5.1   Single-loop second-order modulator

Because of already demonstrated robustness and design simplicity, the single-loop second-order ΣΔ modulator [Cand85] (Fig. 4.24) is one of the most attractive architectures for industrial applications and, hence, widely used.

The netlist of Fig. 4.25 belongs to one of these modulators. In addition to a signal generator, it includes two integrators with gain equal to 0.5 and a comparator with reference voltages 1.5V All the blocks have been considered real with an associated model. The analyses to be performed are:
a) Distortion as a function of the non-linearity of the integrator DC-gain and dynamic limitations.
b) Thermal noise
c) Influence of the comparator hysteresis

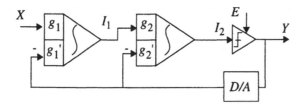

Figure 4.24: Single-loop second-order modulator

```
# Second-order modulator #
Vin inp ampl=0.75 freq=2000;
Comp out (oi2) real cm;
I1 oi1 (inp,out*0.5) real im;
I2 oi2 (oi1,out*0.5) real im;

.output fft(out);
.clock freq=4.096X;
.oversamp 256;
.nsamp 65536;
.options nofdt fullydiff nother;

.model cm Comparator vhigh=1.5 vlow=-1.5 hys=hh;
.model im Integrator cfb=cfbp dcgain=74d ron=1.78k gm=600u imax=30u
cunit=0.25p npwd=5n dcgnl1=5 dcgnl2=dcgnl
 cload=0.1p cpa=2p;

.param hh=sweep(lin 0.0 100m 10m);
.param dcgnl=sweep(lin 0.0 50 5);
.param cfbp=sweep(lin 0.1p 4p 0.1p);
```

Figure 4.25: ASIDES netlist for a second-order $\Sigma\Delta$ modulador

Fig. 4.26 shows some simulation results obtained with the previous netlist. First, the effect of the non-linearity and finite amplifier DC-gain is displayed: harmonic distortion and increase of the quantization noise in the low-frequency range. On the other hand, Fig. 4.27 shows some effects of the thermal noise:

a) comparison of the output spectra with and without such error, and

b) in band noise (+ distortion) power as a function of the value of the sampling capacitor for two sets of integrator dynamic specifications (transconductance and maximum output current). The presence of a minimum demonstrates that an optimum value of the capacitor exists for which the thermal noise contributions KT/C (dominant for low values of the sampling capacitor) and defective settling contribution (relevant for high values of the sampling capacitor) are equal. The position of the minimum depends evidently on the integrator dynamics.

c) Finally, Fig. 4.27(c) shows the response to the comparator hysteresis. Note that hysteresis as high as 10% of the full scale does not degrade significantly the signal-to-noise ratio.

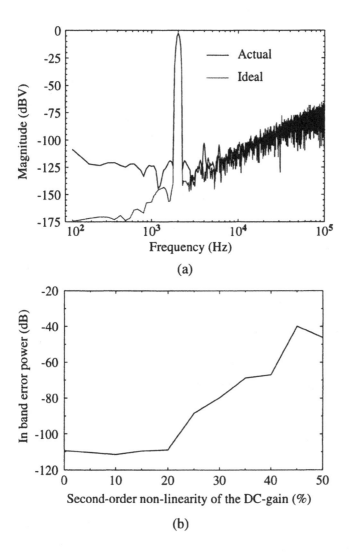

Figure 4.26: Effect of the non-linear amplifier gain: (a) First- and second-order distortion obtained with *dcgnl*1 = 5% and *dcgnl*2 = 20%. (b) In band error power as a function of *dcgnl*2 for *dcgnl*1 = 2%.

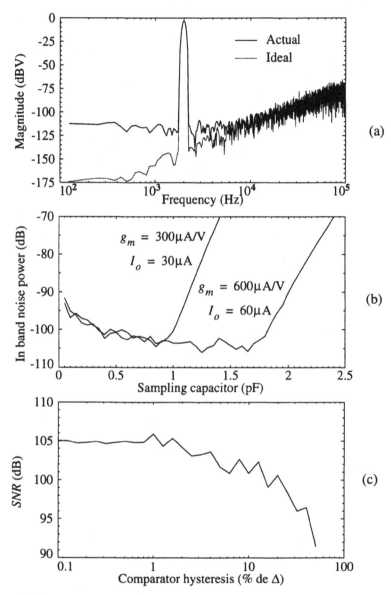

Figure 4.27: (a) Output spectrum with thermal noise. (b) In band noise power as a function of the sampling capacitor for two integrator dynamics. (c) *SNR* in the baseband as a function of the comparator hysteresis. g_m and I_o are the transconductance and the maximum output current, respectively (see Section 4.3.1.2).

4.5.2   Fourth-order 2-1-1 cascade multi-bit ΣΔ modulator

The efficiency of multi-bit quantization in ΣΔ modulators is bound to the solution of the non-linearity problems of the D/A converter in the feedback loop. To that purpose, two well differentiated methods have been followed: (a) calibration (static or dynamic) of the D/A converter elements and use of this in single-loop ΣΔ modulators and (b), use of architectures less sensitive to the errors of the D/A conversion.

The architecture whose simulation is treated in this section, and which will be considered in detail in Chapter 7, constitutes an example of the last trend. The modulator of Fig. 4.28 is formed by the cascade connection of a second-order stage and two first-order stages (2-1-1). Among them, only the last stage includes multi-bit quantization. As will be seen in Chapter 7, the non-linearity errors of the last feedback loop D/A converter are third-order shaped out of the signal band, so that its total power is inversely proportional to the seventh power of the oversampling ratio. Because its principal interest resides in the possibility of reducing the oversampling ratio (using finer quantization), this modulator is especially indicated in medium/high-frequency applications.

The netlist of Fig. 4.29 corresponds to the 2-1-1 cascade multi-bit modulator. The three stages have been clearly isolated. The first two use a comparator as a quantizer, while the final stage includes multi-bit quantization,

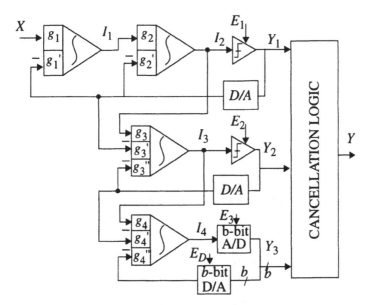

Figure 4.28: 2-1-1 cascade multi-bit ΣΔ modulator

represented here through a quantizer block, and a non-linear block that acts as a D/A converter. The requested analyses are related to the following non-idealities:

```
# SFFCmb: 2-1-1 cascade multi-bit Sigma-Delta modulator#

Vin inp ampl=0.5 freq=125k;

# First Stage
Comp1 out1 (oi2) real cm;
I1 oi1 (inp,out1*0.25) real im;
I2 oi2 (oi1,0*0.5 out1,0*0.25:2) real iota2;

# Second stage
Comp2 out2 (oi3) real cm;
I3 oi3 (oi2,0*1 out1,0*0.375:2 out2,0*0.25:2) real iota2;

# Third stage
Quant_out out3 (oi4) real qm;
Nlb out3fb (out3) real nlb;
I4 oi4 (oi3,0*1 out2,0*0.25:2 out3fb,0*0.25:2) real iota2;

# Cancellation logic

####################

.output snr(out);
.clock freq=32X jitter=200p;
.oversamp 1;
.nsamp 8192;
.options nofdt nother nocfiles mismatch;
.monte = 30;

# Models
.model cm Comparator vhigh=2.0 vlow=-2.0;
.model qm Quantizer maxlevel=2 minlevel=-2 qgain=4 nbits=3;
.model nlb Nlb fsr=4 nbits=3 inl=nl;
.model iota2 Iota2 imax=500u cfb=4p dcgain=72d osp=4
osn=-4 cunit=0.5p gm=3.2m pm=pmo cload=0.6p cpa=1.2p;

.param nl=sweep(lin 0.0 0.1 0.01);
.param pmo = sweep(lin 25 60 1);
```

Figure 4.29: ASIDES netlist of a 2-1-1 cascade multi-bit ΣΔ modulator

a) Mismatching in the integrator gains

b) Clock Jitter

c) Use of small-phase margin amplifiers

To allow the simulation of this last point, a two-pole model has been included for the integrators. Note that the cancellation logic is not shown for the sake of simplicity. This can be found in Section 7.3.2.

Fig. 4.30(a) shows, through a Monte-Carlo analysis, the effect of the mismatching in the capacitor ratios implementing the integrator gains. This type of architecture is especially sensitive to such non-ideality because its operation is based on the cancellation of the first- and second-stage quantization noises, using digital circuitry. The coefficients or weights included in this cir-

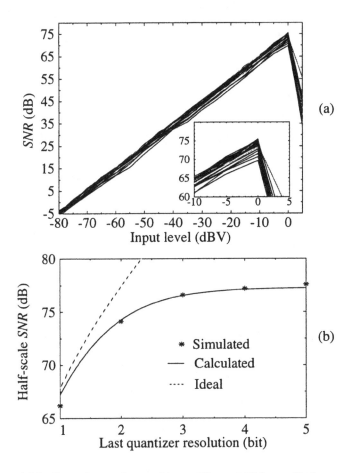

Figure 4.30: Capacitor mismatching effect: (a)Monte-Carlo analysis. (b) *SNR* as a function of the last-stage quantizer resolution.

cuitry must fulfill a relationship with the analog weights, that is affected by the mismatching. Fig. 4.30(b) allows evaluation of the maximum resolution of the last-stage quantizer that leads to a significant increase of the signal-to-noise ratio in the presence of mismatching. The simulated curve has been obtained with ASIDES and corresponds to the worst case of Monte Carlo analysis. The calculated curve has been obtained from the equations of Section 3.2.2. It is observed that values of the last-stage quantizer resolution larger than 3bit do not improve the resolution of the modulator because their effect is masked by the extra noise due to capacitor mismatching.

Fig. 4.31 shows the modulator output spectrum in an ideal case and in the presence of Jitter for two input frequencies. Observe that an important degradation of the noise shaping function is produced (this practically disappears in the larger frequency case) considering only 0.1ns of standard deviation in the clock period.

Fig. 4.32 shows the signal-to-noise ratio as a function of the amplifier phase margin for two sets of dynamic specifications. As stated by the saturation of the SNR curve, in the case A, a phase margin between 45 and 50degree is enough to reach maximum modulator performance.

Finally, Fig. 4.33 shows the influence of the non-linearity of the D/A converter. In spite of the apparent importance of such non-ideality, this architecture presents smaller sensitivity to the errors of the D/A converter than other

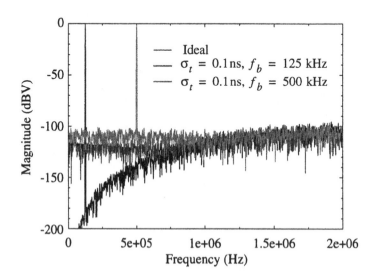

Figure 4.31: Modulator output spectrum with and without Jitter

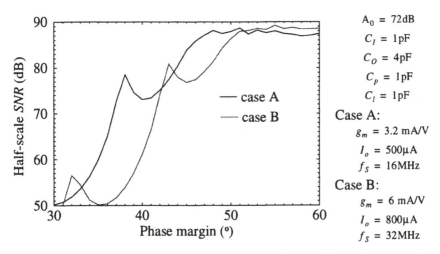

Figure 4.32: Half-scale *SNR* as a function of the amplifier phase margin for two sets of dynamics parameters

cascade topologies and, of course, than the single-loop modulators (see Chapter 7).

SUMMARY

This chapter has been devoted to the simulation of A/D ΣΔ converters using switched-capacitor circuits. The difficulties this task, indispensable as a validating element in the design process, presents are varied: on the one hand, it is not possible to perform electrical simulation of complete modulators because of the length of the transient series needed for their evaluation; additionally, the abstraction of the simulation is risky because, generally, the designer does not have sufficiently accurate models for the basic blocks. This is the fundamental problem of the simulation tools (commercial or not) which we have accessed.

The solution adopted in this work is to develop a tool for the time-domain simulation at a behavioral level of arbitrary architectures of ΣΔ modulators (ASIDES). The simulator interacts with a set of routines written in C language, each of them modeling, as accurately as possible, the behavior of a basic block, with special emphasis on the characteristics most intimately related to their physical implementation. These models, in the form of computational procedures, are based on the analysis of the influence of the circuitry imperfections on the modulator performance carried out in Chapter 3.

As a result, ASIDES permits the simulation of non-ideal effects that are

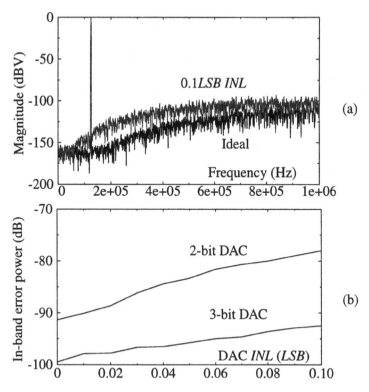

Figure 4.33: Influence of the D/A converter non-linearity: (a) Output spectrum. (b) In-band error power.

not included in related tools, as, for example, the thermal noise, the mismatching, second-order dynamics in the integrators and non-idealities of the D/A converters in multi-bit quantization architectures. The simulator, like the rest of the CAD tools developed, will be used for the design of the $\Sigma\Delta$ modulators in this book.

Chapter **5**

SDOPT+FRIDGE

*Tools for the automatic design of
Sigma-Delta modulators*

5.1   INTRODUCTION

The challenges originated by the design of analog circuits in VLSI technologies, especially conceived for the design of digital circuits, were already pointed out in the introduction of this book [Gray87]:

- The continuous increase in the performance of the digital circuitry requires a parallel evolution of the design techniques for interface analog circuits (basically signal conditioning circuitry and data converters), looking for larger resolution and operation frequency.

- Furthermore, such performance should be obtained in technologies adverse for the analog designer with: (a) continuously decreasing supply voltages, which reduce the dynamic range and, taking into account that the MOS transistor threshold voltage is not equally scaled, practically invalidate certain very popular circuit techniques; and (b) small dimension devices, with poor linearity and matching properties and, frequently, deficiently modeled.

- All this in a noisy environment due to the commutation of the digital circuits, which imposes severe restrictions on the physical design.

These difficulties, together with the intricate relations existing among all the phases of the analog design, from high-level synthesis to the physical level, make the generalization of the interface circuit design intractable. In fact, the traditional solution provided by the analog designers involves, in each particular case, an exhaustive analysis of the systems in order to identify the critical points and the adoption of the architectures, circuits and layout techniques that allow fulfilment of the specifications with the minimum area and power consumption. So, the analysis and synthesis tasks require an important amount of knowledge. For this reason, up to now, the design of interface circuits with low values of the figure $\dfrac{\text{Area} \times \text{Power}}{\text{Resolution} \times \text{Speed}}$ has been accessible only for very experienced designers. Furthermore, said knowledge should be continuously refreshed in order to be adapted to the technology changes and to the market necessities.

This scenario is, to a point, shared by the digital designers. However, these find some advantages: on the one hand, the digital circuits have a consolidated hierarchy – a set of clean-looking rules enables the transfer of the specifications from one level down to the next one. Furthermore, as long as these rules are observed, the functionality of the lower level blocks ensures that of the upper level blocks. From the beginning of the computer era, this characteristic has encouraged the use and diffusion of CAD tools for the analysis, simulation and automation of digital design [Rohr67]. In particular, it is possible to perform the top-down design of a complex digital system, from specifications to silicon, in an automatic way [Bray90], starting from a description of its functionality in a high-level hardware description language like VHDL [Perr93] or VERILOG [Thom91].

Though some of the CAD tools developed for digital design are usable also by the analog designers (like post-layout verification tools), there are hardly any commercial tools specifically conceived to support the synthesis of analog circuits. In fact, a few years ago the only actually available facilities were the electrical simulators, like SPICE [Nage75]. However, electrical simulation, commonly used for the design of basic cells, is clearly inefficient for the verification of more complex systems, due to an excessive consumption of computational resources. This lack is evident when the characterization of the circuits demands long transient analyses, as is the case of the oversampling converters. For these, the electrical simulation of relatively simple circuits, like a first-order $\Sigma\Delta$ modulator may suppose several days of CPU time [Dias92a]. In order to palliate this problem, several tools have been proposed and commercialized, as for instance SWITCAP-2 [Suya90] that enables the simulation of switched-capacitor circuits (SC) with arbitrary clock phases and ELDO [Anac91] and SABER [Saber87] for the simulation of arbitrary mixed-signal circuits. In these tools, the precision, as well as the simulation time are directly related to the quality of the model employed for the basic blocks. If the use of accurate models, within the possibilities of the simulator, is required, the CPU time quickly increases up to several hours for the previous example in a Sparc-10 workstation [Dias92a]. In addition to these commercial tools, multitudes of tools have been published, devoted to a specific set of architectures or covering the specific necessities of their authors. In particular, the behavioral simulation has been profusely used for the design of ΣΔ modulators (see Section 4.1).

Though the simulators are analysis tools and hence valid for bottom-up verification, it is possible to use them for top-down synthesis. This use requires iterating with the simulator at the same time that the design parameters are adequately handled. Thus, to obtain designs in reasonably short times, knowledge is again necessary. This knowledge can be achieved through the simulator, and that is, doubtlessly, another ability of these tools. In any case, the sizing, that is, the mapping of specifications down to design parameter values, iterating with a simulator, results in a very slow process. Because of that, some designers have recently invested considerable effort in the generation of authentic synthesis tools that enable reduction of human intervention in many tasks of the design process.

Traditionally, most of these tools concentrate in the bottom phases of the design cycle: the sizing of basic cells [Degr87][Harj89][Nye88][Ocho94a] [Ocho94b][Ocho96][Youn90][Giel90] and their layout [Cohn91][Conw92] [Kaya88][Rijm89]. Concerning more complex blocks, a very popular strategy is to hierarchize the design; that is, like in digital design, the complex system is partitioned in simpler blocks with relatively independent functionality. In each level of the hierarchy the sizing process involves selecting the architecture and transmitting the specifications to the lower level. This operation must be verified between each couple of levels, generally through simulation, to ensure the correctness of the sizing. The hierarchic synthesis concept is shown in Fig. 5.1[Chan94][Giel96].

Some low-level tools lose their generality as the abstraction level increases. In fact, no tool covering the analog synthesis of mixed-signal blocks of arbitrary complexity and architecture has been reported. Instead of that, synthesis tools have been reported in concrete areas of the analog design like SC filters [Rued91], mostly centered on high-level design, and, thereinafter, for data converters [Giel96] -- more advanced tools covering the design process from the high-level specifications to the physical implementation. The limitation in the number and type of systems is understandable because reliable models for generic complex systems are not available, which makes it necessary to resort to *ad hoc* sizing and validation techniques to establish the synthesis strategy.

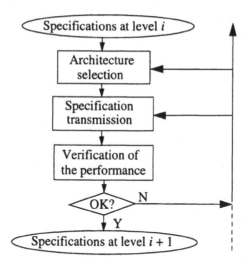

Figure 5.1: Hierarchic synthesis concept [Chan94][Giel96]

Focusing on data converters, Table 5.1 shows a selection of the synthesis tools reported in the last few years. The most important characteristics of the different tools are organized in two blocks corresponding to the two levels of the hierarchy: (a) High-level Synthesis, referring to the sizing at the converter level; that is, the architecture selection and the transmission of specifications down to the subsequent level. Also, the available architectures in each tool, the techniques used for sizing and the verification method are summarized. (b) Low-level Synthesis, that compiles the characteristics related to the basic cell sizing and layout. In addition to the contents in Table 5.1, there are other tools for basic cell synthesis, with which, to the best of our knowledge, no more complex blocks have been sized. The characteristics of some of these tools are shown in Table 5.2. The column CI specifies whether each tool has been used to produce integrated circuits with correct functionality. Finally a column is included with the main advantages and drawbacks of each methodology.

As can be seen in Table 5.1, the synthesis strategies at the system and block level, as well as the verification methods are rather varied. However, all the synthesis strategies can be roughly classified into two large groups:

Table 5.1: Summary of CAD tools for data converter synthesis

Tool	High-level synthesis			Low-level synthesis			CI	Pros / Cons
	Architectures	Sizing	Verification	Basic blocks	Sizing	Layout		
IDAC [Degr87]	1st- and 2nd-order ΣΔMs and digital filters	Fixed design plans based on knowledge	Analytical models	Opamps, comparators, oscillators, voltage references	Fixed design plans based on knowledge	Yes, using ILAC [Rijm89]	Yes	Good accuracy, high speed / Very costly inclusion of new topologies
CADICS [Jusu90] [Jusu92]	Algorithmic ADCs	Design plan + behavioral simulation	Event-driven simulator with C routines	Opamps	Using OPASYN for opamps or LAGER [Brod90] for digital circuits	Yes, using the Univ. of California tools	Yes	Hierarchical design / Sizing of the converter based on simulation. Few architectures, expensively expanded
AZTECA [Hort91]	Successive approximation ADCs and DACs	Optimization based on behavioral simulation	Dedicated functional simulator	The use of IDAC or OPASYM for cell design and ANAGRAM [Garr88] for layout			?	Architecture selection and callibration techniques / Not clear optimization at the basic block level
CATA-LYST [Vita92]	Flash, 2-step flash, subranging and pipelined ADCs	Heuristic + behavioral simulation	Dedicated behavioral simulator	An interface is provided for existing low-level synthesis tools			?	Large number of architectures synthesized from two primitives / Simulation-based optimization very costly for high-level synthesis
MIDAS [Been92]	ΣΔ ADCs and DACs	Based on knowledge	Dedicated behavioral simulator	Hierarchical construction of building blocks from a knowledge data base		Yes, using Philips' tools	Yes	Hierarchical design, proven efficacy / Only covers traditional converter architectures
[Chan94]	Interpolative DACs	Optimization + behavioral simulation	Dedicated behavioral simulator	?	?	Yes, using the Univ. of California tools	Yes	Sizing based on behavioral simulation / Technique not applicable to other data converters

Table 5.1: Summary of CAD tools for data converter synthesis (cont.)

Tool	High-level synthesis			Low-level synthesis			IC	Pros / Cons
	Architectures	Sizing	Verification	Basic blocks	Sizing	Layout		
HiFADiCC [Sabi90]	Successive appro. and pipelined ADCs	Selection of the blocks from a cell library	FIDELDO, HILO	Cell library		Yes, using proprietary tools ALAY and DLAY	?	Interface for standard simulation tools / Optimization reduced to the possibilities of a cell library
CLANS [Kenn88] [Kenn93]	$\Sigma\Delta$ ADCs and DACs	Iterative runs of an analysis routine	Dedicated behavioral simulator	?	?	?	Yes	Tool valid for analysis instead of simulation. / Circuitry imperfections not considered
[Mar95]	2nd- and 3rd-order $\Sigma\Delta$M + digital filters + DSPs	Selection based on knowledge. Sizing based on a cell library	Behavioral simulation	Cell library		Fixed-topology parametrizable blocks.	Yes	Effort put on ASIC-level optimization / Use of fixed, eventually not optimized architectures.

Table 5.2: Summary of CAD tools for basic cell sizing

Tool	Low-level synthesis			IC	Pros / Cons
	Basic block	Sizing	Layout		
OASYS [Harj89]	Opamps	Hierarchical design based on knowledge.	Manual	Yes	Hierarchical design, high speed. New architecture exploration. / Low accuracy, fine-tuning needed.
DELIGHT-SPICE [Nye88]	Arbitrary	Electrical simulation + guided optimization	No	?	Open system, precise final results. / Slower than equation-based methods. Local optimization, pre-tuning of the design parameter needed.
ASTRX/ OBLX [Ocho96]	Arbitrary	AWE simulation + statistical optimization	?	?	Open system. Pre-tuning not needed. Possibility of designing complex cells. / Long execution time. Inaccurate final results.
OPASYN [Youn90]	Opamps	Rule-based architecture selection. Sizing using equations + guided optimization	Yes	?	High speed, useful for exploring design spaces / Closed system, need of new design equations and their derivatives for each new topology, low-accuracy results depending on initial conditions
ARIADNE [Giel92]	Arbitrary	Rule-based selection. Sizing: Equation partially generated by a symbolic analyzer + statistical optimization	Yes	Yes	Partially open system based on equations: high-speed, independent from initial conditions / Difficult to handle complex circuits or with DC specifications.

a) *Knowledge-based Synthesis.* The complexity of the analog synthesis has drawn the attention of the researchers in the field of the artificial intelligence. This activity has resulted in tools like BLADES [El-Tu89] or [Hash89] whose excessive complexity has caused them to have had little success in practical circuit design. However, there are a large number of tools that have inherited and adapted some of the ideas of the artificial intelligence. Such is the case of tools like IDAC [Degr87], CADICS [Jusu92], MIDAS [Been92] and [Mar95] in Table 5.1 or OASYS [Harj89] in Table 5.2. In these, the synthesis strategy is based on capturing the knowledge of experienced designers, either in the form of design heuristic or as approximate equations. The sizing is carried out following a fixed design plan, while heuristic decisions are taken. Although the execution times are very short, the quality of the sizings is not good. Therefore the designs have to be tuned through a simulation tool or through local optimization. Their principal disadvantage is that they are, in general, closed tools; that is, they are limited to a reduced number of topologies and design objectives. Faced with changes in the architecture or in the specifications, the design plans should be remade, which is expensive. An exception is MIDAS [Been92]; in this tool the architectures at the converter level as well as at the block level are obtained by hierarchically merging diverse basic blocks of analog circuits (current mirrors, gain stages, etc.) stored in a base of knowledge. Any analog block (amplifiers, comparators, etc.) composed of the simpler elements of the knowledge database can be synthesized. Because it is not limited to a reduced number of topologies, MIDAS is a less closed tool.

b) *Optimization-based Synthesis.* To this category belong tools such as AZTECA [Hort91], CATALYST [Vita92] and [Chan94] in Table 5.1 and DELIGHT-SPICE [Nye88], ASTRX/OBLX [Ocho96], OPASYN [Youn90] and ARIADNE [Giel92] in Table 5.2. In these tools, the sizing problem is translated in a function minimization problem that can be solved through numerical methods [Aaro56][Rohr67][Teme67]. Previous tools differ in the method employed to evaluate the cost function (function which should be minimized) and in the optimization technique utilized:

With respect to the evaluation of the cost function, a possibility is to use equations as in OPASYN and ARIADNE. The incentive of using equations lies in the computation speed, though the results are always approximate, even more in the case of OPASYN where the equations are obtained manually. Furthermore, in this case the tool is closed, because new equations have to be obtained for each new topology. In spite of using equations, ARIADNE is open, because these are generated automatically by a symbolic analyzer [Giel89]. However, the complexity of the circuit is limited by the possibilities of the analyzer and large-signal (strongly non-linear) specifications cannot be handled [Rodr97]. There also exists the possibility of evaluating the cost function with the help of a behavioral simulator at the converter level, as in AZTECA, CATALYST and [Chan94], or an electrical simulator at the basic cell level, as in DELIGHT-SPICE. This approximation permits better precision in the results, though the execution times are much longer than in the equation-based tools. Furthermore, the capabilities of the simulator determine whether the tool is closed or open. For example, in the basic cell synthesis, the use of electrical simulation, as in DELIGHT-SPICE, extends the possibilities of design to any electronic circuit. Midway between the use of equations and simulation is ASTRX/OBLX where, among others, the AWE (Asymptotic Waveform Evaluation) technique [Pill90][Ragh93] is employed to try to reduce the execution times of the tools based on simulation.

With respect to the optimization techniques, we have approximations based on deterministic methods (DELIGHT-SPICE, OPASYN) and statistic methods (ASTRX/OBLX, ARIADNE). Among the first are the iterative improvement techniques based on gradients. These methods converge quickly but the final solution depends strongly on the initial conditions. That is why they are more adequate for finely adjusting a design obtained manually or by another method. This limitation is solved by the statistic optimization methods, like the simulated annealing technique [Laar87]. This is a very powerful optimization algorithm that permits the global minimization of a multi-variable function at the price of a lower convergence speed. So, it is suitable for the search in large design spaces.

The summary above evidences the difficulty of the automation of the analog design and the large variety of techniques that can be adopted to that end, each one with their advantages and their drawbacks. Focusing on data converters, and especially on those based on ΣΔ modulation, we observe that, though there are some contributions in Table 5.1, these include a reduced set of architectures, with little experimental verification.

In this work we have tried to palliate this situation, developing a set of CAD tools for the automatic design of ΣΔ modulators using switched-capacitor circuits. These tools are vertically integrated to sustain the synthesis process from the high-level specifications to the sizing of the analog cells. For this, as in the other tools of Table 5.1, a hierarchic procedure is followed, based on the concept of Fig. 5.1. The distinctive characteristics of our tools, that will be itemized in the rest of this chapter, are the following:

a) *High-level synthesis*: *Sizing of the modulator* based on equations and statistic optimization; the base of available architectures includes low-order (1st- and 2nd-) modulators, high-order cascade (2-1, 2-2, 2-1-1) modulators, with single-bit or multi-bit quantization. *Verification is* through behavioral simulation.

b) *Low-level Synthesis*: *Basic cell sizing* based on electrical simulation and either statistic or deterministic optimization.

In Section 5.2 the methodology is described and details are given of the different subtools. One of them, the behavioral simulator ASIDES was described in Chapter 4. The remaining tools: SDOPT and FRIDGE, that are used for the sizing of the modulator and basic cells, respectively, will be covered in this chapter. Section 5.3 describes the structure of the equation data base used for the modulator sizing, whose content was itemized in Chapter 3. Sections 5.4 to 5.7 are reserved for the optimizer, a key part of the synthesis process at modulator level as well as at the basic block level. Practical examples of the use of the tools for $\Sigma\Delta$ modulator design are given in Section 5.8 and in Chapters 6 and 7.

5.2   DESCRIPTION OF THE TOOL

Fig. 5.2 shows the operation flow in $\Sigma\Delta$ modulator design. Note that it is an expansion of the hierarchic synthesis concept of Fig. 5.1. The design flow, as for other analog systems, encompasses a set of *synthesis* tasks:

a) *Selection of the topology*, that is, to identify the more adequate modulator architecture for the high-level specifications of the converter (signal bandwidth, resolution, maximum input level, etc.).

b) *Sizing of the modulator*, that is, to find the terminal specifications of the basic blocks that meet the specifications of the modulator.

c) *Selection of the topology of the basic blocks.*

d) *Sizing of the basic blocks*, or mapping of the their specifications in transistor sizes and passive element values.

e) *Generation of the layout of the modulator.*

and *analysis* or verification tasks:

f) *Simulation at the architecture level*, to validate the results of the modulator sizing.

g) *Simulation of the basic blocks* to verify the design of the cells at the electrical level.

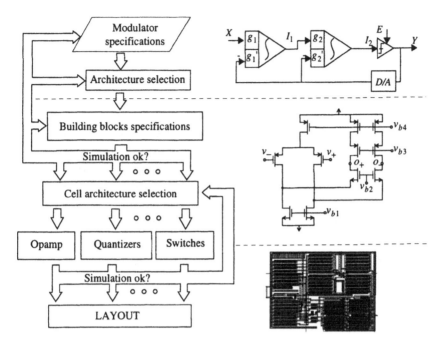

Figure 5.2:   Design flow for ΣΔ modulators

Though not explicitly shown in Fig. 5.2, there is a final verification at the electrical level of the extracted layout of the complete modulator, which is very costly in CPU time and computational resources. Consequently, that simulation covers only a reduced number of clock periods necessary to check the connectivity and to evaluate the degradation of the signals due to layout parasitics.

Among these tasks, the architecture and topology selection requires the attendance of an essential component in the analog designers: knowledge. Because of that, the tools used at this level are intended to facilitate designers with the acquisition of such knowledge [Giel91], relieving them of heavy analysis tasks.

On the other hand, the sizing problem can be transferred to a minimization problem soluble through optimization algorithms. This possibility takes advantage of the growing calculation power of computers to automate the design.

Regarding the analysis tasks, the verification of the basic blocks requires the use of electrical SPICE-like simulators, while that of the modulator is realized much more efficiently using a behavioral simulator, due to the long CPU time consumption needed to iterate electrical simulations at that level (see Chapter 4).

Fig. 5.3 is a flow graph of a vertical integration of the CAD tools that include the following aspects:

a) *Rapid exploration of design spaces* for architecture selection, using simplified design equations.
b) *Sizing of the modulator based on equations.* These equations, studied in Chapter 3, describe the influence of the basic block non-idealities and other architecture features in the modulator performance.
c) *Advanced modulator behavioral simulation.* The behavioral simulator, described in Chapter 4, has been conceived to include arbitrary modulator topologies and all the basic cell non-idealities considered in the equation data-base through more accurate behavioral models.
d) *Sizing of the basic blocks based on electrical simulation.*
e) *Optimum solutions* for the different synthesis tasks thanks to the use of optimization techniques.

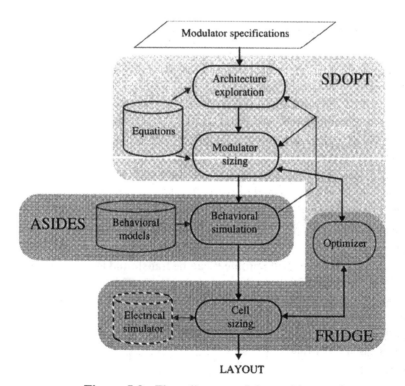

Figure 5.3: Flow diagram of the tool integration

5.3 DESIGN EQUATION DATA BASE

The fast exploration of the design space for the selection of modulator architectures as well as their sizing are based on equations. Among these, the equations that support sizing (see Chapter 3) have been obtained through the exhaustive analysis of various modulator architectures and general switched-capacitor circuits. The equation data base has been conceived to be easily extendable. Thanks to this, the current version includes design equations for a considerable number of modulator architectures, that we can group as follows:

- *Low-order single-loop modulators*, with an order lower than 3 [Cand85][Bose88b][Inos62], which can be made unconditionally stable by properly setting the integrator weights [Op'T93]. For higher-order single-loop architectures, it is not trivial to find a set of integrators that produce stable modulators for any initial condition and input level [Arda87][Wang93]. Several authors provide techniques that permit to obtaining stable modulators with an order larger than 2 [Au97][Adam91][Mous94][Okam93][Rito90]. The diversity of methods used hinders the generalization of the study of these architectures, therefore they are not included in the present version of the data base.

- *High-order cascade modulators*. Contrary to their single-loop counterparts, cascade modulators, also known as MASH (Multi-stage Noise SHaping) [Mats87], do not have instability problems when the order is larger than 2, because they are built through the cascade connection of low-order (< 3) modulators. The architectures of this type considered in the tool are one of third order with the structure 2-1 [Will91][Will94], and two of fourth order, 2-2 [Bahe92][Kare90] and 2-1-1 [Yin93b]. Other architectures with a first-order modulator as a first stage show larger sensitivity to the integrator finite gain and to the capacitor ratio mismatching, so they are not interesting in practice (see Section 3.2.1.1).

- *Multi-bit modulators*. All the above mentioned architectures can be also considered in their multi-bit version, substituting one or several comparators (or single-bit quantizers) by larger resolution quantizers [Teme93].

The equations that describe these architectures can be grouped as follows:
a) Equations related to architectural aspects, that represent the quantization noise power as a function of the non-idealities that affect its shaping function: integrator leakage, mismatching, etc. The analytical expressions that

describe these phenomena depend strongly on the architecture considered and can be rather different for each topology.

b) Equations related to circuit aspects that represent other sources of error different from quantization: thermal noise, incomplete settling error, jitter and harmonic distortion due to non-linearity in the integrators, etc. (see Chapter 3). In all of these, only the contribution of the first integrator to the in-band error power is considered. It is assumed that the contribution of the remaining integrators is negligible because they are attenuated by larger powers of the oversampling ratio. This causes the formulation of these errors to be practically independent of the architecture considered. Nevertheless, the small errors caused by such simplification are evaluated thereinafter during the phase of behavioral simulation.

c) In addition to these careful equations, the tool incorporated a set of simplified equations. These collect aspects related to fundamental limits of the different topologies, envisaging design trade-off in consumption, resolution, speed, etc. They are appropriate for guiding the selection of the modulator architecture as will be shown in Section 5.8.1.

Table 5.3 summarizes the noise and distortion contributions in categories (a) and (b) mentioned above, together with the basic block non-ideality that causes them. The use of these equations for the automatic sizing is demonstrated through practical design cases in Section 5.8 and in Chapters 6 and 7.

Table 5.3: Non-idealities covered by the tool

Building block		Non-ideality	Consequence
Integrators	Opamps	DC-gain finite and non-linear	Quantization noise increase, harmonic distortion.
		Slew-rate	Harmonic distortion.
		Finite GB	Incomplete settling error.
		Limited output swing	Overloading.
		Thermal noise	White noise.
	Switches	Non-zero ON resistance	Settling error, thermal noise.
	Capacitors	Non-linearity, mismatching	Quantization noise increase, harmonic distortion.
Clock		Jitter	Jitter noise.
Comparators		Hysteresis, delay	Quantization noise increase.
Multi-bit quantizers		Non-linearity	Harmonic distortion.

Table 5.4: Approximate expressions of in-band error power for a 2-2 cascade ΣΔ modulator

Quantization noise	$\dfrac{\Delta^2}{12}\left\{\dfrac{4\pi^2}{3M^3}\mu^2 + \dfrac{\delta_A^2\pi^4}{5M^5} + d_1^2\dfrac{\pi^8}{9M^9}\right\}$
Incomplete settling noise	$\dfrac{\Delta^2}{9M}\left(1 + \dfrac{C_p}{C_1}\right)^2 \varsigma^2 \exp\left(-\dfrac{g_m}{C_{eq}}T_S\right)$
Slew-rate harmonic distortion[a]	$\lvert\alpha_3\rvert^2\left(\dfrac{\Delta}{2}\right)^5 /(32k_1^2) + \lvert\alpha_5\rvert^2\left(\dfrac{\Delta}{2}\right)^7 /(512k_1^2)$
Thermal noise	$\left(1 + \dfrac{C_{12}}{C_{11}}\right)\dfrac{kT}{4MC_{11}} + \left(1 + \dfrac{C_{12}^2}{C_{11}^2}\right)\left(\dfrac{kT}{6MC_i} + \dfrac{kTg_mR_{on}}{2MC_i}\right)$
Non-linear capacitor harmonic distortion	$\dfrac{\alpha^2}{8}\left(\dfrac{\Delta}{2}\right)^4 + \dfrac{\beta^2}{32}\left(\dfrac{\Delta}{2}\right)^6$
Non-linear DC-gain harmonic distortion	$\dfrac{\lvert\alpha_1\rvert^2(1+k_1)^2k_2^4}{4A_0^2}\dfrac{1}{k_1^6}\left(\dfrac{\Delta}{2}\right)^4 + \dfrac{\lvert\alpha_2\rvert^2(1+k_1)^2k_2^6}{8A_0^2}\dfrac{1}{k_1^8}\left(\dfrac{\Delta}{2}\right)^6$
Jitter noise	$\left(\dfrac{\Delta}{2}\right)^2\dfrac{(2\pi f_b\sigma_t)^2}{2M}$

a. Complete expressions for α_3 and α_5 are in (3.47).

For example, Table 5.4 collects the approximate expressions of the noise and distortion powers for a 4th-order 2-2 cascade architecture. The symbols used were introduced in Chapter 3.

5.4   CONCEPT AND MOTIVATION OF THE DESIGN SYSTEMS BASED ON OPTIMIZATION

We call sizing to any constructive process that maps *specifications* into *design parameters*. The design specifications should be considered here in a wide sense, including *restrictions* on the performance of a circuit and/or design *objectives*. The meaning of these two terms is clear if we consider, for example, a voltage amplifier whose specifications could be: DC-gain > 70dB, gain-bandwidth product > 5MHz, phase margin > 60degree, input-equivalent noise < 3mV, with minimum power consumption and occupation area. We will call restrictions to the specifications that include inequalities, and objectives to those whose intention is to maximize or to minimize some figure. Observe that the definition of the specifications introduces a character of subordination of the objectives in respect to the restrictions that must be considered in the formulation of the sizing problem.

The sizing of the modulator (or high-level synthesis) as well as the cell sizing, and in general that of any electronic system, can be formulated as a constrained optimization problem. Returning to the previous example, the formulation of this would be:

minimize $pow(\mathbf{x}), area(\mathbf{x})$

$$
\text{subjected to} \begin{cases} A_V(\mathbf{x}) > 70\text{dB} \\ GB(\mathbf{x}) > 5\text{MHz} \\ PM(\mathbf{x}) > 60° \\ N_o(\mathbf{x}) < 3\mu\text{V} \end{cases} \tag{5.1}
$$

where the vector $\mathbf{x}^T = \{x_1, x_2, ..., x_N\}$ represents a point of the multi-dimension space of the design parameters. Note that the constrained character of the formulation is consistent with the degree of subordination expressed previously.

Unfortunately, even in the case of very simple electronic systems or basic blocks, the analytical solution of the sizing problem is not possible due, among other factors, to the following:

- The design equations (mathematical relationships among the specifications and the design parameters) are very difficult to obtain accurately. This is not an excessively serious problem when the degree of abstraction is high, as is the case of the system level descriptions. Note for example the relative simplicity of the equations of Table 5.4. However, using approximated expressions as design equations is dangerous in the basic block synthesis: it is well-known that there is a large disparity between the results of electrical simulation of a simple circuit and those predicted using Level-1 model equations for the MOS transistors in manual design [Alle87].

- The relations between specification and design parameters are typically intricate, and consequently, not solvable analytically, even more if, as usual, the dimension of the design space is large.

- The necessity of minimizing some functions forces the calculation of the first and second derivative introducing additional complications to the analytical solution.

Due to these difficulties, a proper approach to the analog sizing is to use an iterative process. This concept is illustrated in Fig. 5.4: starting from a position x_0 in the design space, a discrete movement sequence is realized until a equilibrium point x^* is reached, which is the solution of the sizing problem. A key part of this process is the calculation of the magnitude and orientation of the movement Δx_n in each iteration. In manual design, choosing this movement is based on the knowledge that the designer has of the system being sized, which constitutes a difficult and laborious task even for the most clever analog designers. The design automation assigns this task to the computer that realizes it according to the evaluation of certain indicatives of the circuit performance.

A convenient approach to this problem is to use a cost function $\Phi(x)$ to quantify the degree of compliance of the restrictions and/or design objectives. Thus, the movement in the nth iteration Δx_n is obtained through functional analysis of $\Phi(x_{n-1})$, $\Phi(x_{n-2})$... This process provides, furthermore, a simple and secure criterion to end the sizing process in points where the cost function is maximized or minimized.

As was explained in Section 5.2, here, the evaluation of the performance reflected in the cost function is realized by equations for modulator sizing and through an electrical simulator for basic block sizing. The use of equations in the first case is justified considering the complexity of the complete characterization at the electrical level of the modulator and the character normally approximate of the high-level representations. On the other hand, it is convenient to use more accurate models, as we approach the physical level,

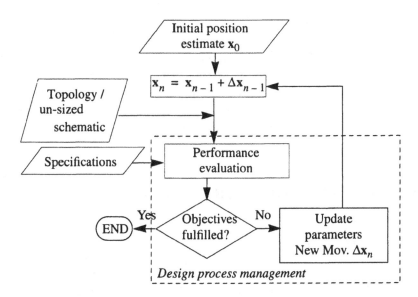

Figure 5.4: General concept of iterative design

to guarantee the feasibility of the circuits, which advises the use of electrical characterization for cell sizing.

Continuing with Fig. 5.4, depending on the formulation of the cost function and on the process updating the design parameters, many possibilities arise for the implementation of the iterative process. In a first approximation, two alternatives can be identified:

- Deterministic incremental techniques, where the calculation of Δx_n requires information on the cost function and on their derivatives [Vand84]. This supposes a great disadvantage because the analytical expression of the cost function is not usually available, so that the derivatives should be calculated via numerical interpolation. Another important disadvantage is that only movements that make the value of the cost function decrease are permitted – the optimization process is quickly trapped in a local minimum of the cost function $\Phi(x)$, so the utility of these techniques concentrates on the fine tuning of sizings obtained manually.

- Statistical techniques, where Δx_n is calculated randomly and hence it does not require information on the derivatives of the cost function [Laar87][Rute89].

The main advantage of the statistical techniques in respect to the deterministic ones is the capability to escape from local minima, thanks to a nonzero probability of accepting movements that increase the cost function. This fact favors the independence of the final solution from the initial position in the design parameter space, which means that it is not necessary to make a manual pre-design. The price to pay is a larger number of iterations and, consequently, larger CPU consumption. However, as will be shown later, a proper formulation of the cost function, and the adjustment of the movement generator to the nature of the analog synthesis palliates the high computational cost and provides a methodology adapted for its automation. These aspects are individually considered next.

5.5   COST FUNCTION FORMULATION

5.5.1   General aspects

The choice of a proper cost function is crucial in an optimization problem. A very powerful algorithm will not work if the function to optimize does not reflect adequately the nature of the problem in question.

In practice, most of the optimization problems are of constrained type [Vand84] and are enunciated like in (4.1). In these, a function depending on a set of design parameters has to be minimized, fulfilling at the same time several restrictions given as inequalities. Such restrictions are simultaneously functions of the design parameters; that is:

$$\text{minimize}[\Phi(\mathbf{x})] \quad \text{sujected to} \tag{5.2}$$
$$\phi_r(\mathbf{x}) \leq 0 \; ; \; 1 \leq r \leq R$$

where R is the total number of restrictions and $\mathbf{x} = (x_1, ..., x_L)^T$ is a position in the design parameter space. The function $\Phi(\mathbf{x})$ is the cost function to minimize, and

$$\phi_r(\mathbf{x}) = y_r(\mathbf{x}) - Y_r \tag{5.3}$$

where y_r stands for a specification and Y_r for its target value. Note that according to the formulation of the restriction in (5.2), the condition $y_r(\mathbf{x}) \leq Y_r$ is pursued. If the contrary condition is wished $y_r(\mathbf{x}) \geq Y_r$ the sign of y_r and Y_r has to be changed.

We will designate *acceptability regions* those within the L-dimension design space where all the restrictions are fulfilled. The constrained optimization problem will consist, hence, of making the system evolve until one of these regions and, once there, finding the optimum of the unconstrained (with no restrictions) problem. Fig. 5.5 is a two-dimensional representation of the design space for constrained optimization. The acceptability regions have been labeled with R_A.

The cost function that is proposed for such a problem is the following:

$$\Psi(\mathbf{x}) = \begin{cases} \Phi(\mathbf{x}) \; if \; \mathbf{x} \in R_A \\ \|\overline{\phi(\mathbf{x})}\|_\infty = max_r[\phi_r(\mathbf{x})] \; if \; \mathbf{x} \notin R_A \end{cases} \tag{5.4}$$

That is, the cost function will take the value of the corresponding cost function associated with the unconstrained problem, if the system is within an acceptability region, and will take the value of the largely unfulfilled restriction, otherwise.

Note that the error function adopted to measure the degree of compliance of the restrictions is the infinite norm; that is, in each iteration only the restriction that is farthest from its target value is considered. This strategy known, as *minimax optimization,* has proven advantageous compared to

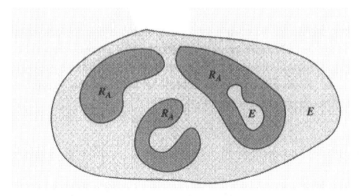

Figure 5.5: Two-dimensional representation of a design space for a constrained optimization problem

other widely used error functions, such as the absolute error (unity norm) or the mean squared error (norm two) [Aaro56]. With these low-order procedures, misleading results can be obtained, as pointed out in Fig. 5.6. In such a figure the absolute error or unity norm defined as:

$$Error = \sum_N |\varepsilon_i| = \sum_N |Y_i - y_i| \qquad (5.5)$$

has been used to measure the deviation of two sets of points from a straight line. In (4.5) N is the total number of points and Y_i, y_i are the actual and ideal values of the i-th ordinate, respectively. Both sets of points have an error equal to 1; however, it is clear that those symbolized by squares fit much better the straight line than those represented by circles. This drawback, graphically shown here, has a clear correspondence with a constrained optimization problem when the number of restrictions is equal or larger than two. The unity norm of the restriction vector $\|\phi(\mathbf{x})\|_1$ can yield very similar values of the cost function for two positions in the parameter space that, however, we would judge very disparate, in particular: (a) all the restrictions are relatively near their targets, and (b) one or a few restrictions are very far from their targets while the remaining (the majority) reach their targets. The situation (a) is clearly advantageous as compared to the situation (b), and (a) is desirable in practice. The appearance of results similar to those of situation (b) is avoidable using larger order procedures: returning to the example of Fig. 5.6, the infinite norm error of the vector corresponding to the circles is still 1, while that of the squares is only 0.2.

So far, the sizing problem has been treated from a generic point of view. What has been said is valid for the modulator sizing as well as for the basic cell sizing. However, it is worthwhile to particularize concerning the formu-

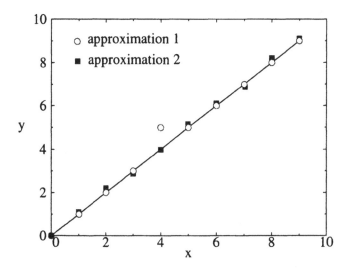

Figure 5.6: Two approximations to a straight line with the same error using the absolute value or unity norm

lation of the cost function.

5.5.2 Cost function formulation for modulator sizing

The high-level synthesis of any electronic system can be envisaged as the following constrained optimization problem: to find the specifications of the basic blocks that ensure the system performance and, simultaneously, are not overspecified in order to facilitate as far as possible their electrical implementation. In particular, for ΣΔ modulators, given a set of high-level specifications (resolution, bandwidth, etc.) the best set of basic block (integrators, amplifiers, comparators, etc.) specifications should be found. Note that in this case there is a single restriction:

$$P_N(\mathbf{x}) - P_{N,max} \leq 0 \tag{5.6}$$

This means: *the in-band error power at the modulator power has to be equal or smaller than the maximum error power permitted to reach the high-level specifications.*

In the acceptability region defined by (5.6), the synthesis problem is reduced to minimize a function whose design parameters are the specifications of the basic blocks. Thus, we can formulate the problem in identical form as in (5.2)

$$\text{minimize}[\Phi(\mathbf{x})] \quad \text{sujected to}$$
$$P_N(\mathbf{x}) - P_{N, max} \leq 0 \tag{5.7}$$

Therefore, the specific formulation of the cost function, with the form given in (5.4) can be:

$$\Psi(\mathbf{x}) = \begin{cases} -\sum_{j=1}^{N} K_j \log\left(\dfrac{x_j}{x_{j, norm}}\right) & si \ \mathbf{x} \in R_A \\[2ex] \log\left(\dfrac{P_N(\mathbf{x})}{P_{N, max}}\right) & si \ \mathbf{x} \notin R_A \end{cases} \tag{5.8}$$

In this expression, logarithms have been used to soften the cost function because the more abrupt a function, the more difficult it is to escape from local minima to reach the global minimum. An example of the benefits of the use of logarithms in the definition of the cost function is shown in Fig. 5.7. The surface of Fig. 5.7(a) has an absolute minimum in $(x, y, z) = (4, 4, 0.5)$, and several local minima with z-axis values very near the absolute minimum. A first sight it seems difficult to locate the global minimum. However, such location is simpler in the representation of the same surface shown in Fig. 5.7(b) where the use of logarithms has signalled the small differences existing between the minima, at the same time as it has softened the function far from the position of the global minimum.

Observe that the restriction in (5.7) remains reflected in the second expression of (5.8) because the logarithm decreases as $P_N(\mathbf{x})$ approaches the maximum permitted error power. The minimum value of the cost function outside of the acceptability region is zero by construction because for error power values lower than the maximum permitted, the system enters an acceptability region and the definition of the cost function does not depend on the obtained error power anymore.

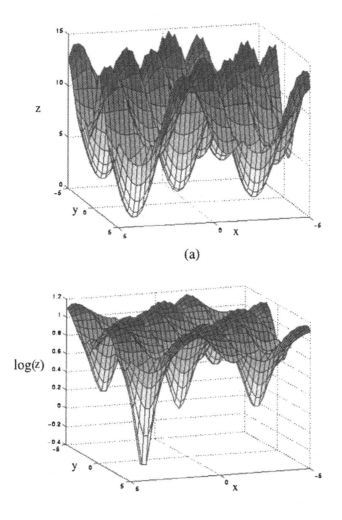

(a)

Figure 5.7: Using logarithms in order to facilitate the localization of the global minimum

Within the acceptability region a sum extended to the total number of design objectives has been chosen as a cost function. The objectives consist of maximizing or minimizing the terminal specifications of the basic blocks (that are simultaneously the design parameters). In each contribution, x_j represents the value of the j-th terminal specification. The contributions are preceded of a weight K_j whose module is used to give more importance to the optimization of some objectives and its sign indicates the tendency of the optimization (positive to maximize, negative to minimize).

Since the absolute values of the block terminal specifications can be rather different, a normalization factor $x_{j,norm}$ is used, defined as follows:

$$
x_{j,norm} = \begin{cases} x_{j,min} & si \ K_j > 0 \\ x_{j,max} & si \ K_j < 0 \end{cases}
\tag{5.9}
$$

where $x_{j,min}$ and $x_{j,max}$ are the lower and upper limits of the variable x_j, respectively.

5.5.3   Cost function formulation for basic block sizing

A peculiar characteristic of the cell-level synthesis is that it usually has a larger number of specifications. For better suitability to the characteristics of the cell design, we propose to classify these specifications as follows:

- *Strong restrictions* or specifications whose fulfillment is considered essential by the designer. A typical example is the feedback stability condition of amplifiers, $PM > 0$ [Alle87]. Non-fulfillment should not be permitted at all on this type of restriction. Any values of the design parameters that do not fulfil one of such restrictions must be, hence, automatically discarded.

- *Weak restrictions*; corresponding to typical specifications of the analog blocks. Remember the example of the voltage amplifier in Section 5.4, equation (5.1). Unlike the strong restrictions, this type of specification allows a certain degree of non-fulfillment, so that the sizings that slightly violate such restrictions can be accepted during the sizing process.

- *Design objectives*; referred to desirable secondary objectives once all the strong and weak restrictions have been fulfilled. As for instance, to minimize the power consumption, to minimize the occupation area, etc.

Note that including the previous specifications in a given category is intended only to illustrate what would be a typical design case. Thus, for example, a designer, though it is not usual, may consider that the power consumption of a circuit must inexcusably be smaller than a certain value, with which this specification would become a strong restriction.

Mathematically, the fulfillment of these three groups of specifications can be formulated as a constrained optimization problem with multiple restrictions and objectives:

$$\text{minimize} \qquad y_{oi}(\mathbf{x}) \qquad , 1 \le i \le P$$

$$\text{sujected to} \quad \begin{cases} y_{sj}(\mathbf{x}) \ge Y_{sj} & \text{or} \quad y_{sj}(\mathbf{x}) \le Y_{sj} & , 1 \le j \le Q \\ y_{wk}(\mathbf{x}) \ge Y_{wk} & \text{or} \quad y_{wk}(\mathbf{x}) \le Y_{wk} & , 1 \le k \le R \end{cases} \qquad (5.10)$$

where y_{oi} stands for the value of the i-th design objective; y_{sj} and y_{wk} are the values of the restriction-type specifications (the subscripts s and w denote the category of the restriction: strong or weak, respectively); and finally Y_{sj} and Y_{wk} are the targeted values of such specifications. Note the correspondence of (5.10) with the generic expression in (5.2).

The cost function can be defined in minimax sense as follows:

$$\Psi(\mathbf{x}) = \begin{cases} \Phi(y_{oi}) & \text{if } \mathbf{x} \in R_A \\ max[F_{sj}(y_{sj}), F_{wk}(y_{wk})] & \text{if } \mathbf{x} \notin R_A \end{cases} \qquad (5.11)$$

where the partial cost functions $\Phi(.), F_{sj}(.)$ y $F_{wk}(.)$ are the following:

$$\Phi(y_{oi}) = -\sum_i w_i \log(|y_{oi}|)$$

$$F_{sj}(y_{sj}) = K_{sj}(y_{sj}, Y_{sj}) \qquad (5.12)$$

$$F_{wk}(y_{wk}) = -w_k \log\left(\frac{y_{wk}}{Y_{wk}}\right)$$

being w_i, the weight associated with the i-th design objective, a positive real number (alternatively negative) if y_{oi} is positive (alternatively negative), $K_{sj}(.)$ is

$$K_{sj}(y_{sj}, Y_{sj}) = \begin{cases} -\infty & \text{if strong restriction holds} \\ \infty & \text{otherwise} \end{cases} \qquad (5.13)$$

and w_k is the weight associated with the weak restrictions that must be a real positive (alternatively negative) if the restriction is of type $\ge$ (alternatively $\le$). These weights are used to give priority to the fulfillment of their associated weak restrictions. In the case of the design objectives, a larger module of the weight will improve the final value of the objective associated to the detriment of the others. As the cost function shows, only the relative magnitude of the weights belonging to the same type of specifications makes sense.

Classifying the circuit specifications leads to a partitioning of the design parameter space. The two-dimensional representation of Fig. 5.5 also applies for cell sizing. In such a figure, the region labeled with E delimits the effective design space that consists of the sub-set of the total space in which all the strong restrictions are fulfilled. Any set of design parameters outside of this region does not represent a useful design; so the actual design space is reduced to such effective space. The acceptability regions, where all the weak restrictions are met, is a sub-set of the effective design space. The regions and sub-regions can be unconnected, which hinders even more the goal of the sizing process: to find the set of design parameters that, within an acceptability region, optimizes the value of the design objectives.

To handle the different types of specifications, the proposed cost function formulation involves the following mechanism:

- In each iteration the fulfillment of the strong restrictions is first checked. If any of them is not fulfilled, the movement that has lead to such a position is automatically rejected.

- The weak restrictions have priority over the design objects. If any weak restrictions are not fulfilled, only that which is farthest from its target is considered to evaluate the cost function. Thus, if the sizing process does not achieve the fulfillment of all the weak restrictions, the final result will be as close as possible to the requirements.

- Once all the strong and weak restrictions are fulfilled, the design objectives are evaluated. The cost function is then a weighted summation of their values.

5.6   OPTIMIZATION ALGORITHM: TECHNIQUES FOR IMPROVING CONVERGENCE

Usually, a larger number of specifications depending on a wide set of design parameters lead to abrupt, non-monotonic cost functions, with lots of local maxima and minima, which hinders the search of the global minimum. This problem becomes much more evident in the basic block sizing where the number of specifications is typically larger and the relations between specifications and design parameters are intricate. Consider for example the cost function of Fig. 5.8. This was obtained applying the formulation of Section 5.5 to a four-transistor differential amplifier, with only two weak restrictions: bandwidth and DC-gain and one design objective: to minimize the power consumption. The design parameters were the width of the input transistors and the biasing current. Note the existence of two acceptability regions (marked with arrows) in the three-dimensional representation of Fig. 5.8(a), where the weak restrictions are fulfilled. They have the aspect of two *valleys* separated by a *mountain chain* where the value of the cost function grows jerkily. Furthermore, as demonstrated in the contour map of Fig. 5.8(b) the two valleys have well differentiated attraction basins, so that it is relatively easy, using a guided, deterministic optimization algorithm, to reach a valley departing from a point placed in its attraction basin. In this example, the valley on the left is deeper than that on the right. In terms of the cost function, this indicates that the power consumption is lower in the first than in the second and the former is, hence, the global minimum being pursued.

However, once the initial conditions are selected within the attraction basin of a minimum, a guided algorithm will surely converge toward that minimum with independence of the fact that it is the global minimum or just a local minimum. Through this example, one of the fundamental problems of the deterministic optimization algorithms in their application to the sizing problems is pointed out: the dependency on the initial conditions.

This inadequacy of the deterministic algorithms for the search in large design spaces, establishes the necessity of using statistical optimization techniques that guarantee independence from the initial conditions, at the price of slower convergence. Between these statistical algorithms, the simulated

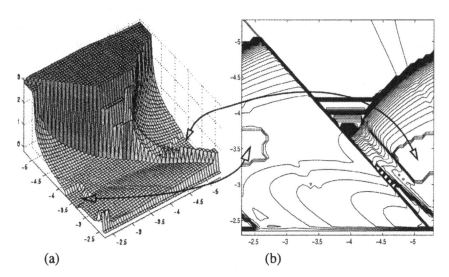

(a)   (b)

Figure 5.8:   (a) Three-dimensional representation of a cost function corresponding to a differential-pair amplifier with 2 specifications and 2 design parameters. (b) Contour map.

annealing technique [Kirk83] is very popular because of its easy implementation, robustness and proven efficacy for the solution of multi-variable optimization problems [Laar87].

Furthermore, as will be seen later, the combination of statistical optimization techniques for the search in a vast design space, and deterministic techniques for the tune-up of the parameters, once the attraction basin of a good minimal is located, permits obtaining admissible sizings with reduced CPU consumption.

The bases of the simulated annealing algorithms are shown in this section and empirical techniques to accelerate the convergence of the previously reported algorithms are introduced. The diverse heuristic techniques will be comparatively evaluated in Section 5.7.

5.6.1   Principles of the simulated annealing algorithms

The optimization algorithm used for the minimization of the cost function consists of a particular case of the simulated annealing algorithm. This algorithm introduced originally by Kirkpatrick et al. [Kirk83] is a general optimization technique for the solution of combinatorial problems whose main features are: (a) random generation of the movements in the design space, (b) probabilistic acceptance of movements that increase the cost function, and (c) independence of the initial conditions. Its name is based on the analogy between the simulation of the annealing of solids and the solution of complex combinatorial optimization problems [Laar87]. In solid-state physics, *annealing* denotes the process by which the temperature of a solid is increased abruptly until its liquid state, wherein the particles of the solid are randomly dispersed. Next, the system is cooled very slowly so that, if the initial temperature was high enough and the cooling was carried out quasi-statically, the particles are ordered, reaching a minimum energy state. To simulate the evolution of the solid until the equilibrium at a given temperature, Metropolis et al. [Metr53] proposed the following algorithm:

a) Generate a small disturbance in the solid through a random movement of one of its particles.

b) If the energy increase ΔE, as a consequence of this disturbance, is negative, the present state is accepted.

c) If, on the contrary, the disturbance has raised the energy, the new state is accepted with a probability

$$P = \exp\left(-\frac{\Delta E}{k_B T}\right) \qquad (5.14)$$

where k_B is the Boltzman constant and T the absolute temperature. Note that the higher the temperature of the solid, the greater the probability of accepting increases in the energy.

The repetition of the Metropolis algorithm for a given temperature leads to the thermal equilibrium of the system in which the energy distribution function approaches a Boltzman distribution [Laar87]. The simulated annealing technique is a reiterated application of the Metropolis algorithm while decreasing the temperature and with it the probability of accepting configurations with increased energetic content. When the temperature is low enough there will not be any possibility of accepting such configurations, which indicates the end of the process.

The Metropolis algorithm can be used to generate the configurations of a combinatorial optimization problem, where the configurations play the role of the energetic states of the solid; the cost function, or function to minimize, that of the energy; and the parameter that controls the probability of accepting increases of the cost function is called, by analogy, the *temperature* of the problem.

Fig. 5.9 shows the block diagram corresponding to a generic simulated annealing algorithm. The shaded region corresponds to the successive application of the Metropolis algorithm on the set of design parameters. The detection of equilibrium breaks this loop also called a Markov chain, and the temperature is updated to continue the process. The decision of accepting or rejecting a positive increase of the cost function is made by comparing its probability, equivalent to that shown in (5.14), with a number randomly generated in the range (0,1). The random nature of the disturbances, together with the statistical acceptance of those that lead to a positive increase in the cost function, make possible escape from local minima and, consequently, the exploration of large regions of the design space. This simple description involves complex mathematics, whose treatment is not necessary in the context of this book. The mathematical aspects of the basic algorithms and others more elaborate are analyzed in detail in [Laar87].

It can be shown [Laar87] that the basic algorithm presented converges with probability equal to 1 to the global minimum if one of the two following conditions is satisfied:

a) For each value of temperature, T_n, an infinite number of movements is generated (which supposes a Markov chain of infinite length) and $\lim_{n \to \infty} T_n = 0$. This possibility is known as *homogeneous algorithm*.

b) For each value of temperature only one movement is generated and T_n evolves to zero slower than $O(|\log n|^{-1})$, or equitably the temperature

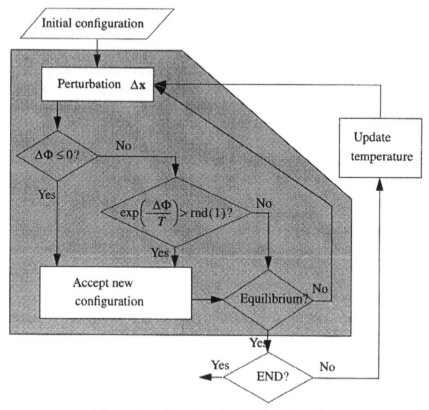

Figure 5.9: Simulated annealing algorithm

reaches zero only after an infinite number of steps; called *heterogeneous algorithm*.

Such conditions are not achievable in practice so that the absolute convergence (with probability 1) is not possible. The closer the temperature evolution or the Markov chain length to one of the previous conditions, the better the approximations. In other words: good results can be obtained either generating many movements for a quickly decreasing temperature (approximation to the homogeneous algorithm), or generating very short Markov chains with a slow decreasing temperature (approximation to the heterogeneous algorithm). The most general case will include a combination of both approximations; that is, a Markov chain of fixed or variable length but finite, and a temperature that arrives at zero after a finite number of steps. In this case, the adequate choice of a set of control parameters can considerably increase the efficacy of the algorithm. These parameters are:

- Initial and final values of the temperature
- Length of the Markov chain
- Temperature updating

The specification of this set of parameters (together with others, for more complex algorithms) constitutes a *cooling schedule*. In literature there are as many cooling schedules as authors [Laar87]. Some of these are based on theoretical calculations, while others arise from purely empirical analysis, sometimes contradicting, the conclusions derived from theory. Table 5.5 summarizes the most significant trends [Laar87].

Table 5.5: Trends in cooling schedules [Laar87]

Parameter	Trends
Initial value of temperature	High enough to accept almost all transitions.
Final value of temperature. (Stop criterion)	Number of iterations determined a priori.
	Stop if the last configuration of a Markov chain does not change in several consecutive chains.
	Stop if the acceptance ratio is below a certain value.
Markov chain length	Fixed, proportional to the number of variables.
	Variable, till a minimum number of transitions are accepted.
	Variable, till a minimum number of transitions are rejected.
	Variable, according to the fluctuations of the cost function during the last chain.
Temperature update law	Exponential decrease.
	Linear decrease.
	History-based variable decrease.

Our optimizer incorporates heuristic improvements regarding the generation of configurations and the cooling schedule. Said innovations have been introduced in order to adapt the classic algorithms to the specific formulation of the cost function, on the one hand, and to the philosophy of the integrated circuit design, on the other. Rather than pursuing the optimum (or global minimum) of a given design with prohibitive CPU time consumption, designs are searched for that fulfill the restrictions with admissible values of the design objectives in reasonably short CPU time. Furthermore, a cooling schedule is proposed, with a simple adaptive mechanism for temperature updating, whose convergence is not significantly degraded as the number of design parameters increases.

The block diagram of Fig. 5.10 represents the operations flow in the optimizer. Also shown is the iteration with the equation data-base for modulator sizing or with an electrical simulator for cell sizing, in order to evaluate the cost function. The content of the shaded blocks, which coincides with those operations introducing innovations, is analyzed in the following sections.

5.6.2 Cooling schedule

The term *cooling schedule* designates the strategy followed in the cooling process that, in classic algorithms, is perfectly defined by four parameters:

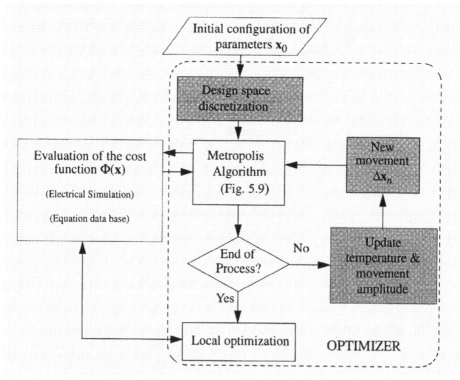

Figure 5.10: Operation flow of the methodology adopted

the initial value of the temperature, the finalization criterion, the temperature update law and the length of the Markov chain. The tool permits the implementation of several cooling schedules among those found in literature. Some of them have been briefly described in the previous section. For example, it is possible to fix the length of the Markov chain, or to make it vary as a function of the history of the process, and it is also possible to fix the initial temperature or to make it proportional to the number of variables, etc. In addition to the classic algorithms, evolved heuristic algorithms are implemented. In this section we will describe adaptations of such algorithms to the particular characteristics of the design problem and innovations regarding the temperature updating law.

As opposed to the classic algorithms, where T decreases monotonously, our tool uses a composed temperature parameter:

$$T(n, \mathbf{x}) = \alpha(\mathbf{x})T_o(n) \tag{5.15}$$

where n is the ordinal of the current iteration, $T_o(n)$ is the normalized temperature that varies with the number of iterations and $\alpha(\mathbf{x})$ is a modulation coefficient that is a function of the position in the design space. Furthermore, a heuristic is incorporated for the selection of $T_o(n)$ and $\alpha(\mathbf{x})$ in order to increase the convergence speed:

- Non-monotonous, adaptive normalized temperature
- Non-linear modulation coefficient with different expressions for the different regions of the design space.

5.6.2.1 Normalized temperature

The normalized temperature, as defined above, is equivalent to the control parameter or temperature of the classic algorithms. Most such algorithms implement a monotonous updating law for the temperature, which supposes that this decreases continually starting from an initial value $T_o(0)$, either exponentially or linearly[1]:

Exponential decreasing: $T_o(n) = \beta_e T_o(n-1) \qquad 0 < \beta_e < 1$

Linear decreasing: $T_o(n) = \dfrac{N-n}{N}T_o(0) \qquad n = 1, ..., N$ (5.16)

Instead of a monotonously decreasing temperature, the use of a cool-

---

1. In [Laar87] several more elaborated updating laws are compiled although they also lead to monotonously decreasing temperatures.

ing/re-heating sequence is proposed. This sequence can be of two types:

a) *Forced* heatings sequence; that is, once the system reaches a low enough temperature, this is increased abruptly until a level that serves as an initial value for a new cooling. Fig. 5.11(a) shows an example corresponding to a set of eight heatings forced with exponential decrease. The control parameters of such an algorithm are the initial and final temperatures, the number of heatings, the decrease law and ratio, etc.

b) *Natural* heatings sequence, where the heatings and subsequent coolings are spontaneous as a consequence of the adaptive temperature mechanism. The algorithm to adapt the temperature is the following:

$$T_o(n) = T_o(n-1) + \gamma\left[1 - \frac{\rho(n)}{\rho_s(n)}\right] \tag{5.17}$$

where ρ, called acceptance ratio, is calculated as

$$\rho(n) = \rho(n-1)\left(1 - \frac{\beta}{\tau}\right) + \frac{\beta}{\tau}a(n) \tag{5.18}$$

where β and τ are constant and $a(n)$ equals 1 if the n-th iteration has been accepted and 0 if it has been rejected. $\rho_s(n)$ is a predetermined acceptance ratio that can be constant or can vary according to certain laws. Fig. 5.11(b) shows the evolution of the temperature if this algorithm is applied in a practical case of cell design[1] with $\beta = 0.9$; $\tau = 50$ y $\rho(0) = 0.8$ and ρ_s decreasing from the initial value of 0.8 until its final value 0.1. The evolution of the acceptance ratio is also shown. Note that the adaptation of the temperature allows us to obviate the correct specification of its initial and final values; because, if eventually too small an initial value is included, so that even slight increases in the cost function are rejected, the temperature will begin to grow quickly. In this sense, this technique resembles those which fix the initial temperature according to the average of the increases in cost throughout a given number of iterations [Kirk83][John87] with the advantage that in this case, if the initial temperature is correct, these previous iterations are not wasted. The CPU time saving can be significant in cell sizing where each iteration may require several electrical simulations.

Our experience demonstrates that this technique, thanks to its relative insensitivity to a wrong choice of the control parameters, obtains the best results in cell sizing and hence this is the algorithm adopted by the default in the tool. In general, the heatings, forced or not, followed by rapid coolings

---

1. Amplifier in Section 6.3.2.1

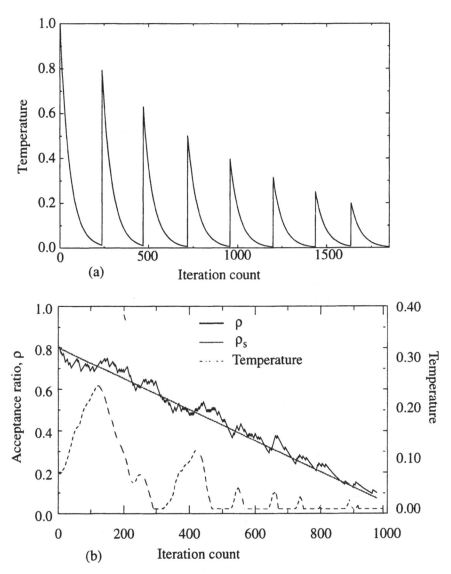

Figure 5.11: Cooling examples: (a) Exponential decrease with reheat-
ings; (b) Adaptive temperature.

procure valid designs in surprisingly short CPU time, when the specifications
of the modulator or cell are not excessively demanding. In cases with very
demanding specifications, we have checked a reduction by a factor of 6 in the
total number of iterations, in comparison with monotonous temperature tech-

niques.

5.6.2.2 Temperature modulation coefficient

As stated in connection with the expression (5.15), the modulation coefficient of the temperature is a function of the region of the design parameter space reached after each movement. The motive of such dependency is to compensate the differences that eventually may appear in the magnitude of the cost function increases in regions of different type. Thus, it is not necessary to define the temperature out of the effective design space, that is, where at least one strong restriction is not met, because any movement that leads to a position out of the effective design space is automatically rejected. On the other hand, in the regions where some weak restriction is violated the temperature is

$$\alpha(\mathbf{x}) = |w_{max}| \Rightarrow T(n, \mathbf{x}) = T_o(n)|w_{max}| \tag{5.19}$$

where w_{max} is the weight associated with the maximum partial cost function among $F_{wk}(.)$ in (5.11) for the current position $\mathbf{x}$, and $T_o(n)$ is the normalized temperature of the n-th iteration. Finally, if the strong restrictions as well as the weak ones are fulfilled, the temperature is defined as

$$\alpha(\mathbf{x}) = \sum_{i=1}^{B} |w_i| \Rightarrow T(n, \mathbf{x}) = T_o(n) \sum_{i=1}^{B} |w_i| \tag{5.20}$$

where w_i is the weight associated with the i-th design objective.

5.6.3   Design parameter updating

The update of the design parameters, or generation of the movements, constitutes another important aspect of the algorithm. In this sense our optimizer presents the characteristics described below:

a) *Amplitude of the* $\Delta\mathbf{x}_n$ *movement varying with the temperature*

The theory indicates that to evolve toward a new configuration it is appropriate to generate small disturbances in the current configuration, which translates into small variations of a certain design parameter. However, our experience with the minimization of complex analytical functions (see Section 5.7) shows that the adoption of a small and fixed amplitude of the random movements is effective only when the temperature decreases very slowly. Otherwise, the system is easily trapped in a local minimum, frequently very far from the global minimum.

Our optimizer incorporates this heuristic considering a variable amplitude for the random movements. In particular, when the temperature is high, large movements are permitted because the increases in cost generated by these will most probably be accepted. Thus, in the beginning, the extensive exploration of the design space is favored. On the other hand, as the temperature decreases, the probability of accepting movements decreases as well and consequently only small movements are realized, which equals fine tuning the design. The variable amplitude of the random movements is obtained using a parameter, ς, for the generation of the n-th position as follows:

$$x_j(n) = x_j(n-1) + \text{rnd}[\varsigma(n)](x_{j,\,max} - x_{j,\,min}) \qquad j = 1, ..., L \qquad (5.21)$$

where $\varsigma(n) = \beta_\varsigma \varsigma(n-1)$, $(0 < \beta_\varsigma < 1)$ and $\text{rnd}(x) = $ random number in $(0, 1)$. Thus, if we make $\varsigma(0) = 1$ in the beginning of the process, the random movements will have an amplitude equal to the full-scale range of each variable $x_{j,\,max} - x_{j,\,min}$, and decrease with the iteration count. ς decreases in all the algorithms implemented in the tool, independently of the technique used to update the temperature.

b) *Possibility of defining logarithmic scales for the design parameters*

Many design parameters have very large variation ranges, especially when the designer does not know, a priori, their approximate values. For cell design good examples of this are the transistor sizes, or the biasing current, whose variation ranges can comprise several decades. In such a case, a change of, for example, $2\mu A$ in a biasing current does not have the same significance for a previous value of this current of $5\mu A$ as for $100\mu A$; linear movements in this variable will cause the low current range to be left partially unexplored. The use of logarithmic scales helps to correct this problem. Thus, the movement generation given in (5.21), becomes

$$x_j(n) = x_j(n-1)\left(\frac{x_{j,\,max}}{x_{j,\,min}}\right)^{\text{rnd}[\varsigma(n)]} \qquad (5.22)$$

Note that, either with linear (5.21) or logarithmic (5.22) movements, it can happen that the new value of the design parameter is out of its range. In that case, said value must be limited to the extremes of the variation range.

c) *Design space discretization*

Many design parameters are of discrete nature; for example, in many IC technologies the transistor sizes can vary only according to a grid. The discretization proposed here is going beyond this because it includes the continuous nature design parameters, and simultaneously defines a larger grid for those of discrete nature. Thus, the design space is conceived as an assembly of hypercubes. Fig. 5.12 is a possible discretization of a three-dimensional design space. Each time a certain hypercube is visited during the optimization process, the value of the cost function at such a point is associated by extension to the rest of the hypervolume. Thus, if in the future a configuration is generated inside an already visited hypercube (see Fig. 5.12), the simulation has not to be performed, so that an important number of simulations are saved with the consequent acceleration of the optimization process.

Note that the validity of associating the same value of the cost function to a region more or less ample of the design space is subordinated to the fact that the function does not present large variations in such region, which is not known a priori. To palliate this drawback it is possible to modify the granularity of the discretization, so that a trade-off between the quality of the minimum reached and calculation speed can be established. However, if with finer discretization the cost function still has significant variations within a hypercube, it would mean that the position of the minimum is very sensitive to small variations of the parameters and the resulting design may not be interesting in terms of physical implementation.

When the optimization process ends, a local optimization is initiated within the hypervolume with the smallest associated cost function, in order

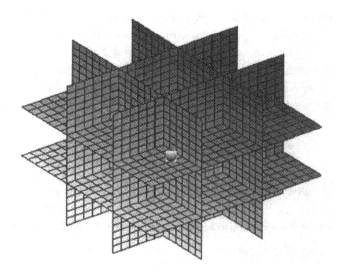

Figure 5.12: Discretization of a three-dimensional design space

to fine-tune the design. In this local optimization the design parameters recover the continuous nature or, alternatively, adapt to the original grid. As previously mentioned, the use of a guided algorithm is recommended in this final optimization for rapid convergence toward the minimum within the hypercube. For that purpose our optimizer implements a variant of the Powell method [Bren70] that does not need direct information from the derivatives of the cost function.

5.7   COMPARISON OF THE HEURISTICS

We will demonstrate the advantages of the proposed heuristics using a multi-minima analytical function whose expression for a N-dimensional case is

$$f(x) = K \tag{5.23}$$

$$\cdot \min\left\{ -e^{-\xi\sum_{k=1}^{N}(x_k - d)^2} \prod_{k=1}^{N}\cos(x_k - d), \quad -e^{-\xi\sum_{k=1}^{N}(x_k - d)^2} \prod_{k=1}^{N}\cos(x_k + d) + \gamma \right\}$$

where K, ξ, d and γ are constant. This function has an absolute minimum of value $-K$ and many local minima, and exhibits the interesting property that the number of minima rises linearly with the number of variables. This means that the complexity of the optimization process is determined exclusively by the number of variables, and not by structural changes in the cost function produced when this number rises. Fig. 5.13 shows this function for two independent variables and the following values of the parameters: $K = 10$, $\xi = 0.002$ $d = 7$ $\gamma = 0.25$.

The different heuristics have been tested using the function in (5.23) for several numbers of independent variables. The test procedure consisted of 100 runnings of each heuristic on the function being optimized, with arbitrary initial conditions in the range of the independent variables ($-20 \le x_j \le 20$, $j = 1, 2...N$) and with an approximately constant iteration count, equal to 2000. For each one of these runs the best achieved minimum was stored. The results of this test are shown in the three-dimensional plots of Fig. 5.14. There, the x axis corresponds to the number of independent variables (from 1 to 11).

The y axis represents the magnitude of the reached minimum; for easier understanding of the results, only integer values of the cost function are represented, so that the minimum reached in each run is represented by its closest integer. The z axis represents the percentage of iterations in which such minimum was reached. The cooling schedule followed in each case is specified briefly in the table attached to each graph.

According to the results obtained, the techniques have been ordered from worse to better in Fig. 5.14.

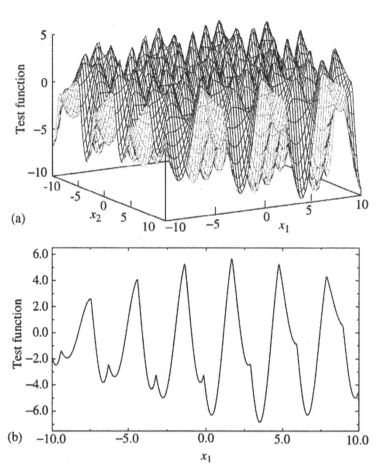

Figure 5.13: (a) Test function for heuristics comparison, (b) Section in $x_2 = 0$.

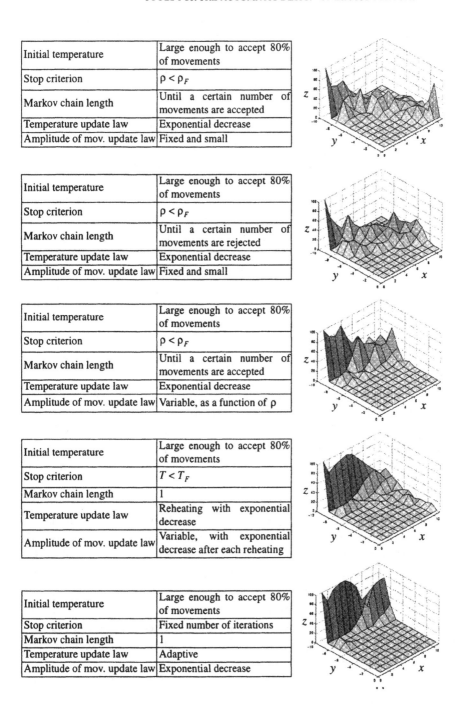

Initial temperature	Large enough to accept 80% of movements
Stop criterion	$\rho < \rho_F$
Markov chain length	Until a certain number of movements are accepted
Temperature update law	Exponential decrease
Amplitude of mov. update law	Fixed and small

Initial temperature	Large enough to accept 80% of movements
Stop criterion	$\rho < \rho_F$
Markov chain length	Until a certain number of movements are rejected
Temperature update law	Exponential decrease
Amplitude of mov. update law	Fixed and small

Initial temperature	Large enough to accept 80% of movements
Stop criterion	$\rho < \rho_F$
Markov chain length	Until a certain number of movements are accepted
Temperature update law	Exponential decrease
Amplitude of mov. update law	Variable, as a function of ρ

Initial temperature	Large enough to accept 80% of movements
Stop criterion	$T < T_F$
Markov chain length	1
Temperature update law	Reheating with exponential decrease
Amplitude of mov. update law	Variable, with exponential decrease after each reheating

Initial temperature	Large enough to accept 80% of movements
Stop criterion	Fixed number of iterations
Markov chain length	1
Temperature update law	Adaptive
Amplitude of mov. update law	Exponential decrease

Figure 5.14: Comparison of cooling schedules

Note that, in all cases, when the number of variables is small, practically 100% of the runnings lead to the global minimum whose value is -10. However, as the number of variables increases (shift to the right in the x axis), and hence the complexity of the problem grows, a dispersion of the function is observed in the first three plots, due to the fact that the quality of the minimum obtained worsens on average. This, though in a lower degree, also happens in the fourth plot; in this case the percentage of success (the fact of reaching either the global minimum or the closest local minimum) decreases quickly when the complexity exceeds eight independent variables. The best result is obtained in the last plot corresponding to the adaptive temperature algorithm described previously. Note that the percentage of success remains close to 80% even in the extreme case with 11 independent variables. It is true that in this case the minimum obtained is not the global (whose value is -10) but one very close to it. This quasi independence of the number of iterations required (remember that it was fixed to 2000 in all cases) of the problem complexity, results in a worthy characteristic in agreement with the philosophy of the electronic design, as stated in Section 5.1.

5.8   DESIGN EXAMPLE: A 16bit at 9.6kS/s $\Sigma\Delta$ modulator

The previous set of tools is used in this section to design a 16bit at 9.6kS/s (4.8kHz signal band) $\Sigma\Delta$ modulator. These specifications are typically required for energy metering applications. The design will be faced with the objective of minimizing power consumption. As stated in this chapter, the achievement of such an objective is intimately related to the following considerations:

a) Selection of the modulator architecture that allows us to obtain the high-level specifications, taking into account the trade-off between complexity of the circuit and required sampling frequency. Note that the benefit, with respect to power consumption, that is obtained using ratio high-order modulators and/or multi-bit quantization with low oversampling ratio, can be masked by the necessary hardware increment.

b) Search for the less demanding specifications for the basic blocks that do not degrade the performance of the modulator. Special attention must be paid to the dynamic requirements of the amplifiers, which usually consume more than 80% of the total power.

c) Selection of proper topologies for the basic blocks and optimum sizing of them.

d) Full-custom layout of the modulator observing the criteria for interference reduction [Tsiv96].

All these premises are intended to obtain modulators with a low value of the figure of merit (*FOM*) introduced in Section 2.4.1, which, for ΣΔ modulators, is defined as follows,

$$FOM = \frac{Power(\text{W})}{2^{resolution(\text{bit})} \times DOR(\text{S/s})} \times 10^{12}$$

where *resolution* stands for the effective resolution of the modulator and *DOR* is its digital output ratio, which equals twice the signal bandwidth, f_d.

5.8.1 Architecture selection

The first step toward the implementation of the modulator consists of selecting the architecture that can fulfil the specifications with minimum FOM. As stated in Section 5.3, at this level a set of very simplified equations is handled to evaluate the power consumption of available modulator architectures. We will consider only architectures with single-bit quantization[†1].

To obtain such equations, we suppose that the dominant error sources in the modulator are quantization, thermal noise and incomplete settling error. Thus, for *b*-bit resolution, the dynamic range (*DR*) of the modulator can be estimated as

$$DR = 3 \cdot 2^{2b-1} = \frac{V_r^2/2}{P_Q + P_{Th} + P_{St}} \qquad (5.24)$$

where V_r represents the full-scale input range of the modulator (which coincides with the reference voltage) and P_Q, P_{Th}, and P_{St} are the in-band powers of the quantization error, thermal noise and of incomplete settling error, respectively. We consider that the incomplete settling error can be controlled by design so that $P_{St} \ll P_Q, P_{Th}$; that is,

$$DR \cong \frac{V_r^2/2}{P_Q + P_{Th}} \qquad (5.25)$$

In this case, using approximated expressions for the quantization and thermal noise, the total noise power in (5.25) depends on only three design parameters: the order of the modulator L, the oversampling ratio M, and the

---

1. The benefits of using multi-bit quantization, recently demonstrated in [Nys96], are difficult to evaluate generically due to the diversity of techniques used to attenuate the non-linearity of the D/A converter [Chen95][Nys96][Leun92] [Sarh93].

first integrator sampling capacitor C_i, as follows,

$$P_Q \cong \frac{(2V_r)^2}{12} \frac{\pi^{2L}}{(2L+1)M^{2L+1}}$$

$$P_{Th} \cong \frac{kT}{MC_i}$$

(5.26)

Using (5.25) and (5.26) it is possible to calculate the minimum value of the capacitor C_i required to obtain given DR as a function of M and L. Once such a value is known, the equivalent load of the first integrator amplifier is evaluated as

$$C_{eq} \cong C_i + C_p + C_l \left(1 + \frac{C_i + C_p}{C_o} \right)$$

(5.27)

where it has been supposed that the SC integrator is like that of Fig. 5.15(a), so that its integration configuration is that of Fig. 5.15(b). The amplifier input parasitic C_p and the integrator output load C_l can be estimated as a fraction ζ of the sampling capacitor. Also the integration capacitor C_o is related to C_i through the integrator gain. Assuming that this gain equals 0.5, (5.27) yields

$$C_{eq} = (1+\zeta)C_i + \zeta C_i \left[1 + \frac{C_i(1+\zeta)}{2C_i} \right] = \left(1 + 2.5\zeta + \frac{\zeta^2}{2} \right) C_i$$

(5.28)

where it has been considered that C_p, $C_l = \zeta C_i$.

To be consequent with the supposition $P_{St} \ll P_Q$, P_{Th}, the gain-bandwidth product of the amplifier, approximated by $g_m/(2\pi C_{eq})$, must be large enough to provide a good settling of the voltages at the integrator output. A

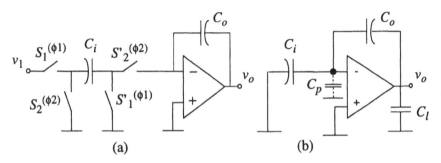

(a)   (b)

Figure 5.15: (a) SC integrator. (b) Integration phase configuration.

conservative choice is

$$g_m/(2\pi C_{eq}) = 5f_S \tag{5.29}$$

where f_S is the sampling frequency. This expression allows an estimation of the transconductance needed in the amplifiers.

At this point one more assumption is necessary concerning the topology of the amplifier. We consider that such an amplifier is a folded-cascode OTA (Fig. 4.5) with the same current I_B flowing in the differential pair and in the output branches; and that the same current is also used in the biasing stage. This way, the total current for the amplifier is $4I_B$. The current I_B depends on the transconductance as

$$I_B = g_m^2/(2\beta) \tag{5.30}$$

where β is the transconductance parameter of the input transistors. Once the current in each amplifier is known, the static power consumption can be approximated as

$$P_{WS} = 4I_B V_{supply} L \tag{5.31}$$

where V_{supply} represents the supply range and L the modulator order.

On the other hand, the dynamic power dissipated to commute a capacitor of value C_i at a ratio f_S between the reference voltages can be estimated as $P_w = (2V_r)^2 C_i f_S$. Using fully-differential circuitry there exist 6 commuting capacitors per integrator (supposing that $C_o = 2C_i$). So, the dynamic power of the analog part of the modulator results in:

$$P_{WD, analog} = 6L(2V_r)^2 C_i f_S \tag{5.32}$$

In addition to this dynamic consumption, the dynamic power dissipated in the digital part of the modulator (quantizers, flip-flops and gates) must be taken into account. However, it is difficult to estimate this consumption because it depends significantly on the number of quantizers in the modulator, as well as on the specific circuitry used in their implementation. A coarse approximation can be

$$P_{WD, digital} = 10Q \cdot 5\text{mW}(1\text{ns})f_S \tag{5.33}$$

where Q represents the number of quantizers (latch + flip-flop + small logic), each one with 10 equivalent inverters, commuting in 1ns with power peaks of 5mW. The value of Q depends on the architecture on the modulator: it is 1 for single-loop modulators or several for cascade modulators. We will suppose that $Q = L/2$, with L being the modulator order.

Using these equations, the FOM of several modulator architectures has been estimated to obtain 16bit at 9.6kS/s. Fig. 5.16 shows the results as a function of the oversampling ratio. The best FOM is obtained with a second-order modulator (Fig. 5.17), with an oversampling ratio close to 300. This is due to the fact that, for such high resolutions, the output spectrum is dominated by unshaped thermal noise. So, though the quantization noise could be reduced by increasing the modulator order, the value of the sampling capacitor cannot, due to the thermal noise restriction, resulting in the same current per amplifier. In such a case, the use of high-order modulators generally leads to larger power consumption.

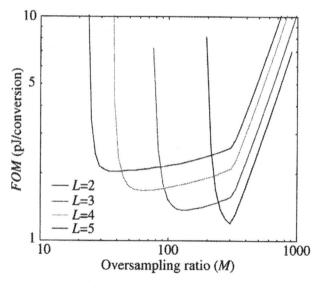

Figure 5.16: Estimated *FOM* to obtain 16bit at 9.6kS/s as a function of the oversampling ratio, using several modulator architectures

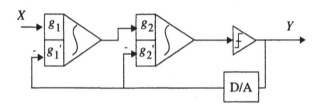

Figure 5.17: Block diagram of a second-order ΣΔ modulator

5.8.2 SC Implementation

The second order modulator has been implemented using switched-capacitor techniques. In addition to the well-known robustness of these techniques, the use of fully-differential circuitry provides the adequate cancellation of the common mode interferences. Fig. 5.18 shows the SC schematic of the modulator. The weights of both integrators have been selected to minimize the needed output range and the dynamic requirements of the amplifiers: using $g_1 = g_1' = g_2' = 0.25$ and $g_2 = 0.5$ the necessary output range coincides with the value of the reference voltages, while using the classic option $g_1 = g_1' = g_2 = g_2' = 0.5$ [Bose88b] such a range must be at least twice the reference voltages. The details on the integrator weight optimization are given in Chapter 6. Note that the first amplifier of Fig. 5.18 includes a chopper technique to attenuate the influence of its offset and low-frequency noise [Hsie81]. Furthermore, the second integrator has two branches to implement two different weights g_2 and g_2'. This is not necessary in the first integrator where the weights of the input and feedback paths are the same.

The timing of the modulator consists of two non-overlapped clock phases and two slightly delayed versions of them, used to avoid the signal-dependent charge injection [Lee85]. These clock phases together with the phase that controls the chopper are shown in Fig. 5.19. The analog switches with large voltage swing, identified by sc in the schematic, are complementary in order to increase the linearity. For these switches, each clock phase must be routed close to its complementary in the layout in order to minimize the substrate noise.

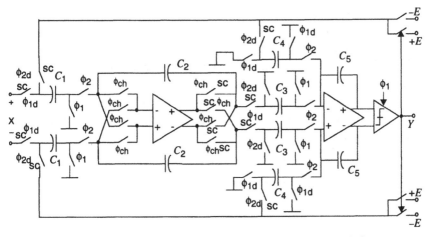

Figure 5.18: Fully-differential SC schematic of the modulator

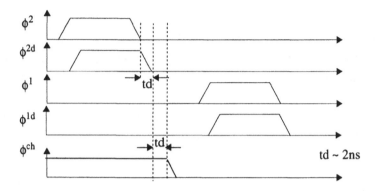

Figure 5.19: Clock phases needed for modulator timing

5.8.2.1 Modulator sizing

The specifications of the basic blocks and other design parameters at the modulator level have been obtained using SDOPT. The input file is shown in Fig. 5.20. The meaning of each parameter is commented on briefly in the figure.

After 2 seconds CPU time the values shown in Table 5.6 were obtained. The reference voltages were fixed to ±1.5V (differential value). The noise and distortion contributions to the in-band error power are shown at the end of Table 5.6. Note that the incomplete settling of the voltage at the integrator output is the main error source – a consequence of taking a very adjusted value for the dynamic parameters of the amplifier (transconductance and output current), in order to minimize the power consumption. Other error sources, including quantization noise, have approximately the same influence. The in-band error power referred to the full-scale input amplitude ($V_r = 1.5$V) gives 98.3dB dynamic range, which is enough to obtain 16-bit effective resolution.

```
specify sd2n {                          // Modulator specifications
        type SL;                        // Topology: Single-loop modulator
        bits 16;
        nyquist band 9.6k;
        refvoltage 1.5;
        order 2;
}
variables {
        M   = 256;                      // Oversampling ratio
        ADC = [1000, 2000, 500]-1.0;    // Opamp DC-gain
        Io = [25u, 50u, 15u]-1.0;       // Maximum output current
        Gm  = [150u, 500u, 50u]-1.0;    // Opamp transconductance
        H   = [50m, 100m, 20m]1.0;      // Comparator hysteresis
        R   = [2k, 5k, 1k]1.0;          // Switches ON resistance
        NLC1 = [20, 50, 1.0]1.0;        // Capacitor non-linearity (ppm/V)
        ADC_NL1 =[0.1,1.0,0.01]1.0;     // 1st-order opamp DC-gain non-linearity (%/V)
        ADC_NL2 =[20,50,1]1.0;          // 2nd-order opamp DC-gain non-linearity (%/V²)
        A = 1.0;                        // Input amplitude for harmonic distortion calculation
        Jitter = [3,10,1].5;            // Std deviation of the clock period (ns.)
        Ci = [1p,3p,1p]-1.0;            // Sampling capacitor
        Cp = 0.2p;                      // Opamp input parasitic
        Cl = 0.1p;                      // Integrator output parasitic
        G1 = 0.25;                      // First integrator weights
        G1p = 0.25;
}
iterations {
    numitera 5000;
}
spscan {

    // Calculation of signal / (noise+distortion) ratio

    print A*A/(2*(PQ+PST+PSR_HD+PTH+PJITTER+PHD_OP))
    PROGRAM     XGRAPH    db;
            A    range  1e-6 2 nstep 10 dec;
}
```

Figure 5.20: Input file for SDOPT, design parameter are briefly commented on.

Table 5.6: SDOPT results

OPTIMIZED SPECS FOR:		16bit@9.6kS/s@1.0V
Modulator	Topology	2n-order
	Sampling frequency (MHz)	2.4576
	Oversampling ratio	256
	Reference voltages (V)	±1.5
Integrators	C_1, C_3 (pF)	1
	C_2 (pF)	4
	C_4 (pF)	0.5
	C_5 (pF)	2
	Capacitor non-linearity (p.p.m.)	≤ 50
	MOS switch-ON resistance (kΩ)	2.0
	Maximum clock jitter (ns)	≤ 2.0
Opamps	DC-gain (dB)	≥ 66
	DC-gain non-linearity (V^{-2})	≤ 20%
	g_m (μA/V)	196
	Maximum output current (μA)	≥ 30
	Total output swing (V)	≥ 4.0
	Input noise density (nV/sqrt(Hz))	≤ 20
	Parasitic input capacitor (pF)	≤ 0.2
Comparators	Hysteresis (mV)	≤ 70
	Resolution time (ns)	≤ 50
RESOLUTION & NOISE POWER CONTRIBUTIONS		
Dynamic range:		98.3dB (16.04bit)
Quantization noise (dB)		-108.7
Thermal noise (dB)		-107.1
Incomplete settling noise (dB)		-99.3
Jitter noise (dB)		-118.1
Harmonic distortion (dB)		-107.0

To validate the sizing of the modulator, behavioral simulations have been realized using ASIDES. Fig. 5.21 shows the simulator input netlist. Observe that each non-ideal block (integrators and comparator) has an associated model that contains the specifications derived from SDOPT. Fig. 5.22(a)

shows the output spectrum of the modulator and Fig. 5.22(b) shows its sig-
nal/(noise + distortion) ratio (*TSNR*) in the baseband as a function of the
input amplitude. Fig. 5.22(b) also shows the *TSNR* obtained with SDOPT.
The good agreement between the simulated and the calculated curve confirms the validity of the sizing of the modulator. As will be seen later, said
agreement is kept with the measurements.

```
# Second-Order Sigma-Delta Modulator #
###################################

Vin inp ampl=0.5 freq=1.25k;          # input signal

Comp out (oi2) real cmi;              # Comparator
I1 oi1 (inp,out*0.25) real im2;       # 1st integrator
I2 oi2 (oi1,0*0.5 out,0*.25:2) real im2;# 2nd integrator

.output fft(out);
.clock freq=2.5X jitter=2.0n;         # clock definition
.oversamp 256;                        # Oversampling ratio
.options nofdt fullydiff;

# Comparator model
.model cmi Comparator vhigh=1.5 vlow=-1.5 hys=70m;

# Integrator model
.model im Integrator
cfb=4p                    # sampling capacitor
ron=2k                    # switch on-resistance
dcgain=66d               # opamp DC-gain
dcgl=20                  # 2nd-order DC-gain non-linearity (%)
gm=196u                  # opamp transconductance
imax=30u                 # opamp max. output current
npwd=20n        # opamp thermal noise spectral density (V/sqrt(Hz))
osp=4                    # opamp output swing (±2V)
cpa=0.2p                 # opamp input parasitic
cload=0.1p               # integrator load capacitor
cunit=0.25p              # unitary capacitor
cnl=50;                  # capacitor non-linearity (ppm)
```

Figure 5.21: ASIDES input netlist for the 2nd-order modulator

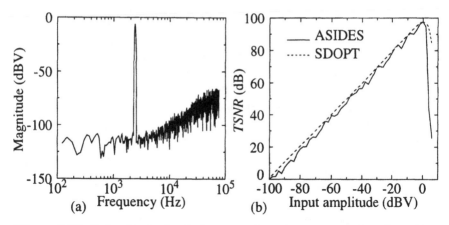

Figure 5.22: Behavioral simulation: (a) Output spectrum for sinusoidal input with amplitude 0.5V and frequency 1.2kHz. (b) Signal/(noise+distortion) ratio vs. input level.

5.8.2.2 Amplifier design

The relatively low DC-gain and output current requirements allow use of a single-stage class-A architecture for the amplifier. In particular, the fully-differential folded-cascode OTA [Ribn84] of Fig. 5.23 has been selected. A NMOS-input topology has been preferred to a PMOS-input one because the flicker noise will not be relevant, provided that the chopper is activated. On the other hand, due to their smaller transconductance for the same dimensions and current, the PMOS transistors generate more thermal noise, which must be included in the total noise power. In Fig. 5.23 the amplifying stage is

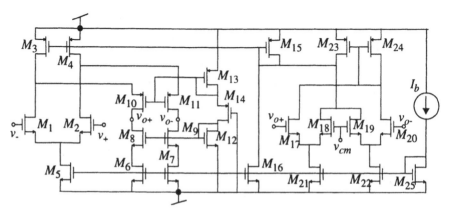

Figure 5.23: Fully-differential folded-cascode amplifier

formed by transistors M_1 to M_{11} while the transistors M_{12} to M_{14} form the biasing stage. The rest of the circuitry, (M_{15}-M_{24}) is a low-distortion common-mode feedback net (CMFB) [Duque93]. Transistor M_5 is used to generate the biasing voltage of the NMOS current sources, while the PMOS current sources are regulated by the CMFB net.

This topology was automatically sized using FRIDGE to meet the specifications of Table 5.6 within temperature and supply ranges of [-25°, 85°] and [4.75V, 5.25V], respectively. After 25min. CPU time, the sizes shown in Table 5.7 were obtained. The biasing current was 30μA. Table 5.8 summarizes the corresponding worst-case simulation results in the temperature and voltage range.

Table 5.7: Amplifier sizes (μm)

Transistor	W/L	Transistor	W/L
$M_{1,2}$	70.5/2	M_{14}	2/4
$M_{3,4}$	15.8/2	M_{15}	15/2
M_5	37.2/2	M_{16}	37.2/2
$M_{6,7}$	20.5/2	$M_{17,18}$	2/4.5
$M_{8,9}$	39.9/1.2	$M_{19,20}$	2/4.5
$M_{10,11}$	78.8/1.2	$M_{21,22}$	37.2/2
M_{12}	5/5	$M_{23,24}$	15/2
M_{13}	5/5	M_{25}	37.2/2

Table 5.8: Simulated results for the amplifier

Specification	Worst-case	Units
Open-loop DC-gain	85.1	dB
Transconductance	196	μA/V
GB (1.2pF)	20.7	MHz
PM (1.2pF)	70	Degree
Input white noise	17	nV/√Hz
Differential ouput swing	5	V
Maximum output current	30	μA
Power consumption	0.63	mW

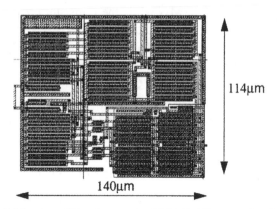

Figure 5.24: Amplifier layout in CMOS 0.7μm technology

Fig. 5.24 shows the layout of the amplifier realized manually in a 0.7μm CMOS technology occupying $140 \times 114 \mu m^2$. All the transistors that should be matched were implemented using common-centroide techniques [Tsiv96]. Furthermore properly biased guard rings were placed to palliate the effect of substrate noise.

5.8.2.3 Comparator design

Since the sensitivity requirement for the comparator is not so demanding while the resolution time is (see Table 5.6), a latched architecture can be used [Yuka85]. The schematic of the comparator is shown in Fig. 5.25. The latch, Fig. 5.25(a), is strobbed in the positive edge of ϕ_1. During phase ϕ_2 data at

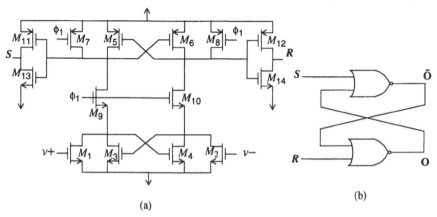

Figure 5.25: Comparator: (a) Regenerative latch. (b) SR flip-flop.

the latch output is accessible through an SR flip-flop made with NOR gates. The latch was sized using FRIDGE to fulfil the specifications of Table 5.6. Results are shown in Table 5.9. It is possible to use very small input transistors because the offset of the comparator is attenuated by the high DC-gain of the integrators in the loop. The simulation results, corresponding to the worst conditions in the temperature and supply voltage range are shown in Table 5.10.

Table 5.9: Comparator sizes (μm)

Transistor	W/L	Transistor	W/L
$M_{1,2}$	1.5/5	$M_{5,6}$	4/2
$M_{3,4}$	1.5/5	$M_{7\text{-}10}$	2.2/0.7
		$M_{11\text{-}14}$	2.2/0.7

Table 5.10:Simulated results for the comparator

Specification	Worst-case	Units
Hysteresis	0	mV
Offset	-10	mV
Resolution time	9	ns

5.8.2.4 Clock phase generator design

Fig. 5.26 shows the schematic of the clock phase generator. The first branch is intended to generate two non-oversampled phases, ϕ_1 and ϕ_2, and two delayed versions of them, ϕ_{1d} and ϕ_{2d}. This branch also generates the corresponding complementary phases. The second branch generates the phases controlling the chopper operation in the first integrator that can be activated or de-activated through the enable signal. The chopper control consists of two non-oversampled phases ϕ_d and ϕ_c, and two slightly delayed versions of them, ϕ_{dd} and ϕ_{cd}. Fig. 5.27(a) shows the form in which the chopper phases control the switches that commute the first integrator input and output. The results of the simulation corresponding to the extracted layout are shown in Fig. 5.27(a).

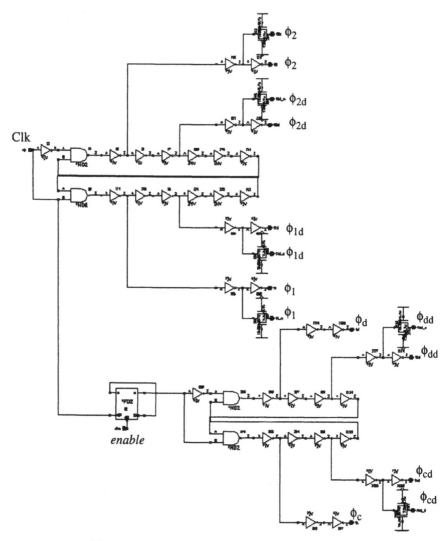

Figure 5.26: Clock phase generator schematic

5.8.3 Experimental results

For test purposes, two chips were fabricated in a CMOS 0.7μm technology. One of them contained the basic blocks of the modulator: amplifier and comparator while the other contained the complete modulator and the clock phase generator. The results measured in both chips are given next.

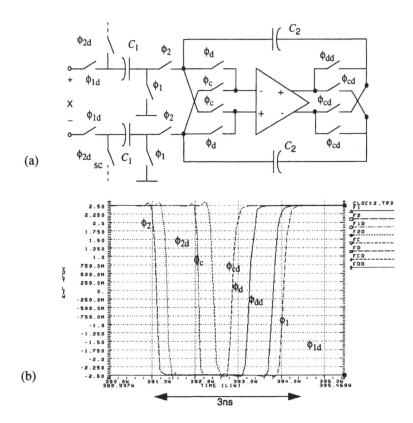

Figure 5.27: (a) First integrator chopper control detail. (b)Clock phase transitions obtained by electrical simulation.

5.8.3.1 Basic blocks

Table 5.11 summarizes the characteristics measured for the amplifier. Results correspond to an eight sample average. The static measurements show that the DC-gain is slightly smaller than expected. However, that is high enough so as not to degrade the modulator performance. Also, a small deviation is observed in the value of the transconductance. These small discrepancies can be due to the fact that the biasing current supplied during the test was slightly larger than the nominal that was used in the simulations. The rest of the characteristics are in agreement with the simulation results.

Table 5.11: Amplifier experimental results

Specification	Measured	Units
Open-loop DC-gain	80	dB
Transconductance	210	µA/V
Maximum output current	positive 29.2	µA
	negative -31	
Differential output swing	5	V
Input offset	3.6	mV

Fig. 5.28 shows the transfer curve of the comparator obtained at 2.5-MHz clock rate. The offset as well as the hysteresis considerably varied from sample to sample. The average and worst-case values of both characteristics, together with the resolution times on eight samples are given in Table 5.12.

Table 5.12: Comparator experimental results

Specification	Average	Worst-case	Units
Hysteresis	28	34	mV
Offset	-27	-35	mV
TP_{HL}	9.4	-	ns
TP_{LH}	10	-	ns

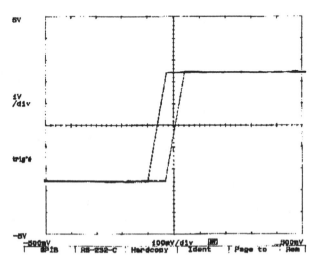

Figure 5.28: Transfer curve measured for the comparator

5.8.3.2 Modulator

Fig. 5.29 shows a microphotograph of the modulator fabricated in a CMOS 0.7μm single-poly technology which occupies 0.42mm^2 and dissipates 1.71mW at 5-V supply. To perform the test of the modulator, a two-layer printed board was fabricated with separate analog and digital ground planes, decoupling capacitors and filtered input signal in order to attenuate the commutation noise in the signal and biasing traces [LaMa92] (see Chapter 6). The modulator input was provided using a differential output high-quality (*THD*<100dB) programmable sinusoidal generator. The acquisition of the modulator output series, synchronized by the clock signal, was realized using a digital circuit characterization unit (HP82000). Data were acquired automatically controlling the measurement set-up with specific routines written in C language. Next, data were transferred to a work station to perform the filtering and decimation using MATLAB [Math91].

Fig. 5.30(a), (b) and (c) show the output spectrum of the modulator when the chopper operation is de-activated for an input tone of frequency 1.25kHz and several amplitudes. The lobes in the low-frequency zone, that reflect the offset of the modulator, are due to the limited resolution of the FFT. Observe that the noise floor (flat in the signal band, except for said lobes) is around -115dBV. Furthermore, there are no significant spurious tones, nor harmonic distortion. Fig. 5.30(d) shows the output spectrum with no input when the chopper operation is activated and de-activated. Note the reduction of the off-

Figure 5.29: Microphotograph of the 2nd-order modulator and clock generator

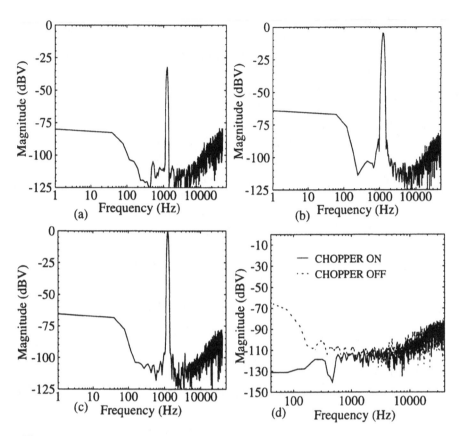

Figure 5.30: Measured output spectrums, with sinusoidal input of amplitude (a) -32dBV, (b) -2dBV and (c) 0dBV and frequency 1.25kHz. (d) Chopper offset compensation.

set by a factor 1000 when the chopper is active.

Measurements of the signal-to-noise ratio (*SNR*) and signal-(noise+distortion) ratio (*TSNR*) were obtained by processing the modulator output with a third-order Sinc filter implemented by software. Fig. 5.31(a) shows both ratios for a varying-amplitude sinusoidal input at 1.25kHz, and oversampling ratio $M = 256$ (nominal value). Fig. 5.31(b) compares the *SNR* obtained for $M = 256$ and $M = 128$. The dynamic range measured for $M = 256$ is above 100dB with an *SNR*-peak of 94.2dB corresponding to a 1.2-V amplitude and a *TSNR*-peak of 91.8dB for 0.8V. In respect of the case M = 128, the dynamic range is 92dB with peaks of *SNR* and *TSNR* of 84.4dB for 1-V input amplitude.

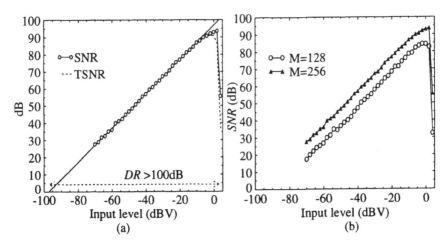

Figure 5.31: (a) *SNR* and *TSNR* measured for an input of frequency
1.25kHz and variable amplitude, with $M = 256$. (b) *SNR* for
$M = 256$ and $M = 128$. In both cases, the clock frequency
was 2.56MHz.

Measurements of the dynamic range as a function of the oversampling
ratio for fixed clock frequency (2.5MHz), Fig. 5.32, show that for $M = 256$
the modulator is just in the limit between the zone where the in-band error
power is dominated by quantization noise (slope = 15dB/octave) and the

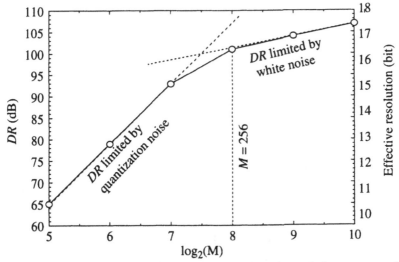

Figure 5.32: Measured dynamic range as a function of the oversampling
ratio at the nominal clock rate (2.56MHz)

zone where it is dominated by white noise (slope = 3dB/octave). This fact is coherent with data in Table 5.6, where it was predicted that the in-band power of the incomplete settling error, which appears as white noise in the signal band (see Chapter 3), was going to be the dominant error source.

Table 5.13 summarizes the measured characteristics for three values of the oversampling ratio. For the nominal value, 256, the modulator presents an *FOM* of only 2pJ/conversion, which is one the lowest in Table 2.4. However, note that the *FOM* increases for the two other values in Table 5.13. Thus, though it is possible to obtain a larger resolution for $M = 512$ or a larger bandwidth for $M = 128$ with the same modulator, the value of the *FOM* rises in either case. Obtaining the minimum *FOM* for the nominal value of the oversampling ratio confirms the efficiency of the methodology presented in this book.

Table 5.13: Summary of the modulator performance

	$M = 128$	$M = 256$	$M = 512$	Units
Effective resolution	15	16.4	17.1	bit
Dynamic range	92	100.2	105	dB
SNR peak	84.4	94.2	101	dB
TSNR peak	84.3	92	99	dB
Digital output rate (*DOR*)	19.2	9.6	4.8	kHz
Maximum input amplitude		1.25		V
Minimum supply voltage		4		V
Consumption (5-V supply) (Sampling rate = 2.5MHz)		1.71		mW
Active area		0.42		mm^2
Figure of merit (*FOM*)	2.7	2.0	2.5	pJ

SUMMARY

The contents of this chapter have been centered on the development of a design methodology based on CAD tools. The tools, SDOPT for modulator sizing and FRIDGE for basic cell sizing, are vertically integrated together with the behavioral simulator ASIDES, to complete the design, from the modulator high-level specifications to the sizing of the basic building blocks, in a very short time.

Furthermore, the incorporation of optimization techniques in the synthesis tasks allows the modulators designed with this methodology to obtain very good values of the figure of merit, $\dfrac{Power(W)}{2^{resolution(bit)} \times DOR(S/s)}$ (*FOM*). This is demonstrated in this chapter through the design of a second-order ΣΔ modulator. The measurements performed on a modulator prototype fabricated in a CMOS 0.7μm technology show an equivalent resolution of 16.4bit at 9.6kS/s with a power consumption of only 1.71mW (*FOM*=2pJ). This performance places the modulator between those with lowest value of the *FOM* reported until now.

Chapter **6**

Integrated circuit design (I)

A 17-bit 40kSample/s fourth-order cascade Sigma-Delta modulator

6.1   INTRODUCTION

Low-order (first- and second-) $\Sigma\Delta$ modulation-based A/D converters [Agra83][Cand85][Inos62][Leun88][Plass78] are very interesting for mixed-signal applications due to their small analog circuitry content, robustness and easy implementation. The second-order architecture [Agra83][Bose88b] [Cand85] is preferred to the simpler first-order one because it is less sensitive to the correlation between the input and the quantization noise, which reduces the presence of pattern noise [Cand81], and has been largely used in various industrial applications. In any case, to reach a given resolution the oversampling ratio (M) must be kept constant, so that an increase in bandwidth implies a proportional increase in the clock frequency. For example, if a 16bit resolution, is required for a second-order architecture, M must equal at least 256, which for a 2-kHz bandwidth (typical in energy metering systems) pushes the clock frequency up to 1.024MHz. In audio band (20kHz), a 10.2MHz clock rate would be needed, while for video, up to 5MHz, 2.56GHz! Apart from the necessary increase in power consumption, increasing the clock frequency generates additional problems such as jitter, switching noise, etc.

The other way to increase the bandwidth (for a given resolution) in $\Sigma\Delta$ converters involves increasing the modulator order. This increases the order of the noise-shaping function, removing a larger amount of error power from the signal band. In Chapter 2, some of the available architectures for high-order modulators were overviewed. Among them, cascade modulators [Long88][Mats87][Rebe90], whose generic block diagram is shown in Fig. 6.1, are especially attractive. As mentioned previously, their operation is

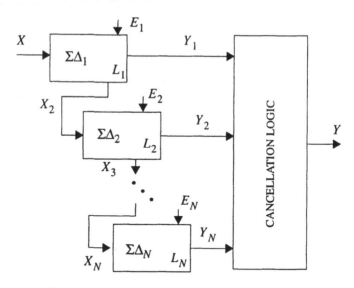

Figure 6.1: Generic cascade $\Sigma\Delta$ modulator

based on the cascade connection of low-order modulators (0, 1 and 2), whose stability is guaranteed by design. The quantization noise generated in a stage (modulator) is re-modulated in the following stage/s and afterwards canceled in the digital domain. As a result, in the ideal case, a delayed version of the modulator input is obtained, plus the last-stage quantization noise attenuated by a shaping function of order equal to the total number of integrators in the cascade. These architectures need an extra digital part to generate the cancellation of the low-order quantization noise, which does not mean a big problem regarding the high degree of automation of today's digital design. On the other hand, the connection in cascade of several stages helps de-correlate the quantization noise and the input, so that the modulators obtained are more robust against the presence of idle tones in the signal band [Cand81][Op'T93].

This chapter is devoted to the study and implementation of cascade $\Sigma\Delta$ modulator architectures. In Section 6.2, two fourth-order topologies with two and three stages are analyzed from an ideal point of view. The imperfections derived from the electrical implementation that affect the behavior of both architectures are analyzed in the end of this section. Section 6.3 illustrates the implementation using switched-capacitor circuits of a two-stage fourth-order modulator. This has been designed using the CAD tools described in Chapter 5. Finally, experimental results are presented in Section 6.4.

6.2   FOURTH-ORDER CASCADE MODULATOR ARCHITECTURES

In Chapter 2, cascade modulator architectures were studied from an ideal viewpoint. The impact in the modulator performance of the building blocks' non-idealities was analyzed in Chapter 3. Now we will use those results to compare two fourth-order cascade architectures: (a) the one formed by two second-order stages (2-2) [Kare90][Bahe92], and (b) the one with a second-order stage and two first-order stages (2-1-1) [Yin93b]. Both architectures are shown in Fig. 6.2. Other fourth-order cascades are of little use in practice

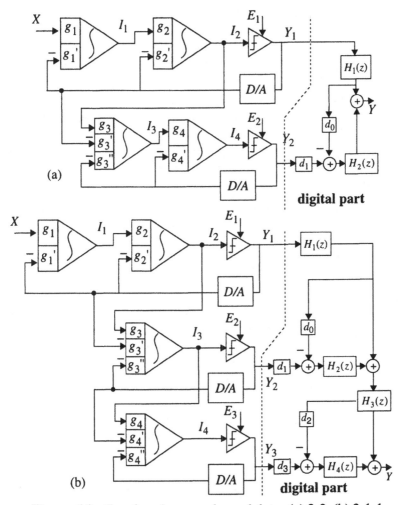

Figure 6.2: Fourth-order cascade modulator (a) 2-2, (b) 2-1-1

because of their larger sensitivity to the integrator leakage, as demonstrated in Chapter 3.

6.2.1 Ideal study

This section is intended to compare the previous architectures from the ideal point of view; that is, taking into account only the quantization error. To that end, optimization principles will be used in order to make a selection of the scale coefficients that maximize the dynamic range in each case.

6.2.1.1 Coefficient relationships

Recall that the cancellation of the first-stage quantization noise imposes certain bounds among the scale coefficients of the analog part (integrator weights) and those of the digital part. As shown in Section 2.3.3, after a generic Z-domain analysis, these relationships are determined by simply demanding that the transfer functions of the quantization noise other than the last be $0^{\dagger 1}$, and that the module of the signal transfer function yield unity. These calculations lead to the relationships of Tables 6.1 and 6.2, valid for the 2-2 and 2-1-1 architectures, respectively.

Table 6.1: Relationships for the 2-2 architecture

Analog	Digital/Analog	Digital
$g_1' = g_1$	$d_0 = 1 - g_3'/(g_1 g_2 g_3)$	$H_1(z) = z^{-1}$
$g_2' = 2g_1'g_2$	$d_1 = g_3''/(g_1 g_2 g_3)$	$H_2(z) = (1 - z^{-1})^2$
$g_4' = 2g_3''g_4$		

Table 6.2: Relationships for the 2-1-1 architecture

Analog	Digital/Analog	Digital
$g_1' = g_1$	$d_0 = 1 - g_3'/(g_1 g_2 g_3)$	$H_1(z) = z^{-1}$
$g_2' = 2g_1'g_2$	$d_1 = g_3''/(g_1 g_2 g_3)$	$H_2(z) = (1 - z^{-1})^2$
$g_4' = g_3''g_4$	$d_2 = \left(1 - \dfrac{g_3'}{g_1 g_2 g_3}\right)\left(1 - \dfrac{g_4'}{g_3''g_4}\right) \equiv 0$	$H_3(z) = z^{-1}$
	$d_3 = g_4''/(g_1 g_2 g_3 g_4)$	$H_4(z) = (1 - z^{-1})^3$

1. The shaping function of the last-stage quantization error must be proportional to $(1 - z^{-1})^4$

With such relationships, the following Z-domain expressions are obtained after the cancellation logic:

$$Y|_{2-2}(z) = X(z)z^{-4} + d_1(1-z^{-1})^4 E_2(z)$$
$$Y|_{2-1-1}(z) = X(z)z^{-4} + d_3(1-z^{-1})^4 E_3(z)$$

(6.1)

where $X(z)$ and $Y(z)$ stand for the Z-transform of the modulator output and input, respectively; $E_2(z)$ and $E_3(z)$ are the Z-transforms of the last-stage quantization noise in each case; and d_1 and d_3 are scalars larger than unity as a result of the necessary scaling of the signals in the analog part. This produces a systematic loss of resolution in respect to the ideal operation. The minimization of such a loss has to be kept in mind during the selection of the coefficient values.

Up to now ideal conditions have been supposed, so that any set of coefficients that fulfil the equations of Tables 6.1 and 6.2 leads to mathematically correct modulators. However, to get a most realistic view, other considerations related to the physical implementation should be considered. Thus, it must be taken into account that:

- The output swing (OS), which depends on the integrator weights as well as on the input level, must be physically achievable. In SC implementations, this limit clearly depends on the supply voltage.

- The levels of the signal transferred between stages should be low enough to avoid premature overload of the following stages.

Unfortunately, the analytical attainment of both figures as a function of the integrator weight is complex and requires a detailed analysis of the time-domain operation of the modulator-stages. This analysis is dealt with in Appendix A. Using the iterative procedures developed in this appendix, it is possible to estimate the OS required in the integrators of each stage. These calculations can be used to optimize the coefficients for each topology. Optimization, in this context, means determining the value of the analog coefficients that, fulfilling the relationships of Tables 6.1 or 6.2, produce:

a) Minimum quantization noise; that is, digital coefficients d_1 and d_3 in (6.1) as small as possible.

b) Easily achievable output swing required in the integrators, taking into account the supply voltage.

c) Transition signal levels that do not prematurely overload the following stage/s. These levels are approximately $\pm V_r$ (reference voltage) for a first-order stage and $\pm 0.9 V_r$ for a second-order stage [Op'T93].

The optimization was carried out using a statistical procedure like that described in Chapter 5, trying to minimize the cost function collecting the previous restrictions. The results for both architectures are shown in Table 6.3.

Table 6.3: Optimized integrator weights

Coefficients	2-2	2-1-1
g_1	0.25	0.25
g_1'	0.25	0.25
g_2	0.5	0.5
g_2'	0.25	0.25
g_3	1	1
g_3'	0.375	0.375
g_3''	0.25	0.25
g_4	0.5	1
g_4'	0.25	0.25
g_4''	-	0.25

With these values, the digital coefficients are:

$$d_0 = -2 \qquad d_1 = 2 \qquad d_2 = 0 \qquad d_3 = 2 \qquad (6.2)$$

and the required OS is $\pm 1.2 V_r$ for the 2-2 modulator and $\pm V_r$ for the 2-1-1 modulator. With these values the Z-domain output of both modulators given in (6.1) yields

$$Y|_{2-2}(z) = X(z)z^{-4} + 2(1 - z^{-1})^4 E_2(z)$$

$$Y|_{2-1-1}(z) = X(z)z^{-4} + 2(1 - z^{-1})^4 E_3(z)$$

$$(6.3)$$

The factor 2 that accompanies the quantization noise is responsible for a systematic loss of 6dB in the dynamic range.

Fig. 6.3 shows the signal-to-noise ratio (SNR) vs. the input level referred to the full-scale (0dB=reference voltage, V_r) obtained with ASIDES (Chapter 4) for $M = 64$. The coefficients of Table 6.3 are considered, together with ideal elements, except for the output ranges of the integrators, which are fixed to the previous values. In respect to the SNR, no important differences

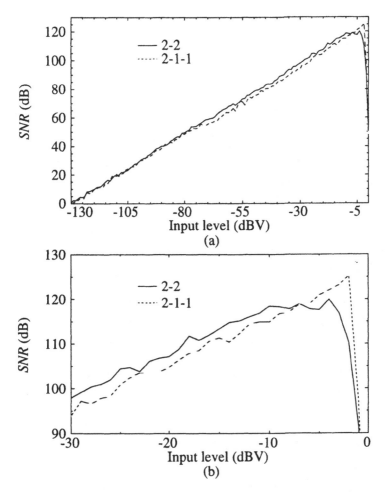

Figure 6.3: (a) *SNR* against input level. (b) Detail of the large input range.

are observed in Fig. 6.3(a) except for the input level close to V_r. This region is shown in detail in Fig. 6.3(b). This difference is because the second stage of the 2-2 modulator is a second-order modulator, whose maximum input level is approximately 10% smaller than that of a first-order modulator, for the same reference voltages. Consequently, the overload of the 2-2 modulator 2-2 is produced slightly before that of 2-1-1 and, at least ideally, the *SNR* peak is 5dB larger for the latter in the case considered.

6.2.2 Non-ideal effects

Up to now, with the exception of the OS limitation of the integrators, the comparison of the architectures has been done considering ideal elements, and the simulations performed correspond to idealized versions of the two topologies. However, a realistic comparison has to take into account the degree of sensitivity of both architectures to the imperfections of the circuitry.

The non-idealities that degrade the modulator performance can be grouped in two categories:

a) Non-idealities whose impact can be modeled as an error source at the first integrator input. This simplification is possible because the contribution of the remaining integrators is at least M times less significant due to their position in the loop. These non-idealities include thermal noise, incomplete settling, non-linear DC-gain, etc. In general, there will be no practical differences between the topologies under study because both have a second-order modulator as a first stage; consequently, the study of such non-idealities, accomplished for generic modulators in Chapter 3, is not necessary in this context.

b) Non-idealities that cause changes in the signal and quantization noise transfer functions. This category includes the well-known sensitivity of cascade architectures to the integrator leakage and weight mismatching [Ribn91]. Both imperfections produce modulator gain error and incorrect cancellation of the quantization noise in the first stage/s. As will be seen later, there are differences between the architectures considered.

The non-idealities belonging to this last category were analyzed in detail in Section 3.2. Here we will simply use the results obtained there in order to compare their impact on both fourth-order architectures.

6.2.2.1 Integrator leakage

In SC implementations, the final expressions for the in-band quantization noise power, including integrator leakage due to finite amplifier DC-gain, are (see Section 3.2.1):

$$P_{Q,2-2} \cong \frac{\Delta^2}{12} \left\{ \frac{4\pi^2}{3M^3}\mu^2 + d_1^2\frac{\pi^8}{9M^9}(1+2\mu) \right\}$$

$$P_{Q,2-1-1} \cong \frac{\Delta^2}{12} \left\{ \frac{4\pi^2}{3M^3}\mu^2 + d_1^2 \cdot \frac{\pi^4\mu^2}{5M^5} + d_3^2\frac{\pi^8}{9M^9}(1+2\mu) \right\}$$

(6.4)

where $\mu = g/A_V$, with g being the integrator weight, A_V stands for the amplifier open-loop DC-gain and M is the oversampling ratio. The difference between both lies in the presence of an extra term in the 2-1-1 modulator

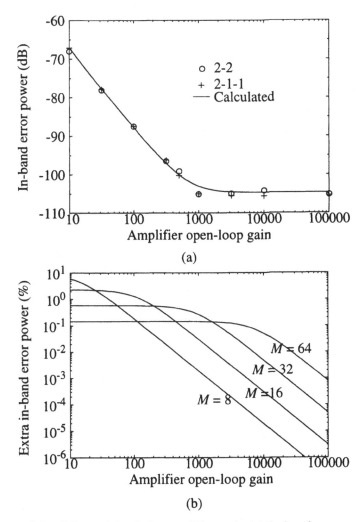

Figure 6.4: Effect of the finite amplifier gain (a) In-band error power for the two architectures with $M = 32$. (b) Excess of in-band error power for the 2-1-1 modulator with respect to that of the 2-2 modulator, as a function of the amplifier gain for several values of M.

power, which is inversely proportional to M^5. For typical values of M (256, 128, 64, etc.), this term is clearly negligible compared to the term divided by M^3. So, there are no differences between the 2-2 and 2-1-1 architectures concerning the sensitivity to the finite DC-gain. This is shown in Fig. 6.4(a) and (b). The former shows the in-band error power obtained by behavioral simulation for $M = 32$ as a function of the amplifier DC-gain. Also, the analytical curves from (6.4) are shown. The last figure shows the extra in-band error power for the 2-1-1 architecture with respect to the 2-2 modulator, as a function of the amplifier DC-gain, for several values of M. Observe that differences exist only for very low values of M and extremely low values of the DC-gain. In a practical case, in the central zone of the plot, there are no appreciable differences.

6.2.2.2 Weight mismatching

In SC implementations, integrator weights are obtained through capacitor ratios. In Section 3.2.2, expressions were developed that relate the in-band error power of cascade $\Sigma\Delta$ modulators to the physical characteristics of the SC implementation, as for instance the size and number of the unitary capacitors, technology features and geometric disposition of the capacitor layout. We repeat here the results obtained then for the approximate increases in quantization error power:

$$\Delta P_Q|_{2\text{-}2}(dB) = 10\log\left[1 + \frac{9}{5d_1^2}\left(\frac{\varepsilon_1 + \varepsilon_{g_1'}}{1 - \varepsilon_1}\right)^2\frac{M^4}{\pi^4}\right]$$

$$\Delta P_Q|_{2\text{-}1\text{-}1}(dB) = 10\log\left[1 + \frac{9}{5d_3^2}\left(\frac{\varepsilon_1 + \varepsilon_{g_1'}}{1 - \varepsilon_3}\right)^2\frac{M^4}{\pi^4}\right]$$

(6.5)

where

$$\varepsilon_1 = \frac{\Delta g_3''}{g_3''} - \frac{\Delta g_1}{g_1} - \frac{\Delta g_2}{g_2}\frac{\Delta g_3}{g_3} \qquad \varepsilon_{g_1'} = \frac{\Delta g_1'}{g_1'}$$

$$\varepsilon_3 = \frac{\Delta g_4''}{g_4''} - \frac{\Delta g_1}{g_1} - \frac{\Delta g_2}{g_2} - \frac{\Delta g_3}{g_3} - \frac{\Delta g_4}{g_4}$$

(6.6)

are the relative errors of d_1, g_1' and d_3, respectively, whose worst-case values can be estimated as 3 times their standard deviation given by

$$\sigma_{g_i} = \frac{n_i}{m_i}\sqrt{\left(\frac{1}{n_i} + \frac{1}{m_i}\right)\left(\frac{K_{le}}{C_u^{3/2}} + \frac{K_{lo}}{C_u}\right) + \frac{2K_{ge}}{C_u} + 2K_{go}}$$

(6.7)

where n_i and m_i stand for the number of unitary capacitors C_u that implement the numerator and denominator respectively of the weight g_i. The constants determine the local variations (with subscript l) and global variations (with subscript g) of the oxide thickness (with subscript o) and of the etch-

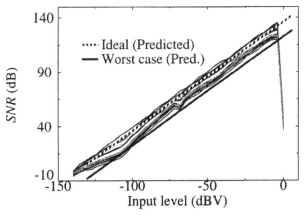

Figure 6.5: *SNR* vs. input level in the presence of mismatching for the 2-2 modulator

ing process (with subscript e) [Shyu84]. Due to the larger number of calculations, it is difficult to perform the comparison analytically. As an alternative we will use the behavioral simulation. Fig. 6.5 shows a group of *SNR* curves obtained using Monte Carlo analysis, where each integrator weight value is assumed to present a Gaussian distribution with a mean equal its nominal value and the standard deviation calculated from (6.7) with

$$K_{le} = 5.8\times10^{-24} \qquad K_{lo} = 1.133\times10^{-18} \qquad K_{ge} = 0 \qquad K_{go} = 0 \quad (6.8)$$

corresponding to a 1.2μm CMOS technology, where the global effects are first-order cancelled by using common centroide techniques in the layout [Tsiv96]. The simulation conditions and the results are summarized in the first column of Table 6.4.

Table 6.4: Simulation conditions and results for the 2-2 and 2-1-1 modulators

	2-2	2-1-1
Sampling frequency	5.12Mhz	5.12Mhz
Oversampling ratio	64	64
Reference voltages	1.0V	1.0V
Unitary capacitor	0.25pF	0.25pF
Integrator weights	See Table 6.3	See Table 6.3
Number of experiments	30	30
SNR peak standard deviation	3.3%	4.6%
Worst case *SNR*	106dB	104dB

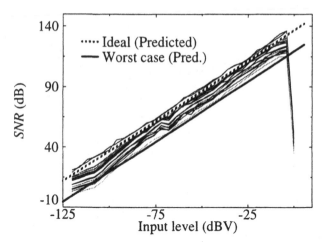

Figure 6.6: *SNR* vs. input level in presence of mismatching for the 2-1-1 modulator

Fig. 6.6 and the second column of Table 6.4 show the equivalent results for the 2-1-1 topology. Note that the latter has a higher sensitivity to the mismatching so that, though in the ideal case the SNR peak for the 2-1-1 modulator exceeds that of the 2-2 modulator in 5dB, the worst-case simulation, taking into account mismatching of the integrator weights, yields for the former an *SNR* peak 2dB smaller than that obtained for the latter.

For this example, both architectures show a considerable degradation of the *SNR* with respect to the ideal case: 16dB in the 2-2 modulator and 21dB in the 2-1-1 one. Let us consider an alternative to both modulators which is the 2-2 modulator with all the analog coefficients equal to 0.5 except $g_3' = 0$. This set of coefficients, that we will call the classic option (2-2/c), because they were proposed along with the architecture [Bahe92], leads, according to the relationships of Table 6.1, to $d_0 = 1, d_1 = 4$. With which the modulator output after digital cancellation is

$$Y|_{2-2/c}(z) = X(z)z^{-4} + 4(1 - z^{-1})^4 E_2(z) \tag{6.9}$$

Fig. 6.7(a) shows the *SNR* obtained by behavioral simulation in the ideal case. Observe that the peak is smaller in 6dB for the 2-2/c modulator. This loss of resolution is because the digital coefficient that amplifies the quantization noise is twice that of the previous architectures. On the other hand, according to the calculations of Appendix A, the *OS* required in the first-stage integrators is $2V_r$, approximately twice that in the optimized cases.

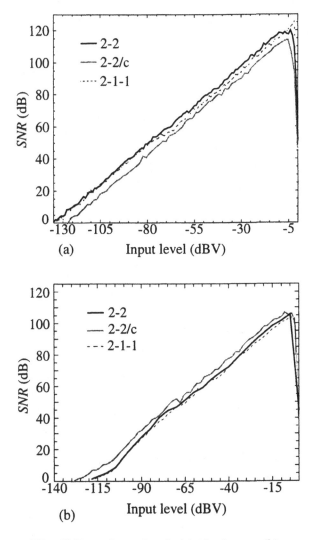

Figure 6.7: *SNR* vs. input level: (a) ideal case, (b) worst case in the presence of mismatching.

In spite of its inferiority in the ideal case, the 2-2/c architecture is advantageous in the presence of weight mismatching. This is illustrated in Fig. 6.7(b), where the worst-case simulations of the *SNR* for the three modulators are shown. For $M = 64$, the worst-case *SNR*-peak, corresponding to the 2-2/c modulator, is 3dB and 5dB higher than those of the 2-2 and 2-1-1 modulators, respectively. The reason for this lower sensitivity to the mismatching is found in the previous analysis: on the one hand, the digital coefficient d_1 in (6.5) is larger in the classic option so that the increase of the quantization noise is smaller; on the other hand, the fact that g_3' is zero in this architecture, allows use of a single-branch integrator as the first integrator of the second stage. With which

$$\frac{\Delta g_3''}{g_3''} = \left(\frac{\Delta C_3''}{C_3''} - \frac{\Delta C_3^o}{C_3^o}\right) = \left(\frac{\Delta C_3}{C_3} - \frac{\Delta C_3^o}{C_3^o}\right) = \frac{\Delta g_3}{g_3} \tag{6.10}$$

because C_3'' and C_3 stand for the same capacitor. Thus, the contributions to ε_1 of g_3 and g_3'' are cancelled (see Section 3.2.2).

As a concluding remark, Table 6.5 collects the distinctive features of the topologies analyzed. Data have been grouped into three categories: from the ideal point of view, taking into account the degree of sensitivity to mismatching, and considering the circuit complexity.

Table 6.5: Comparison results

		2-2/c	2-2	2-1-1
Ideal	*SNR* peak (dB)	117	120	125
Non ideal	Output swing required (*OS*)	±2Vref	±1.2Vref	±Vref
	SNR peak standard deviation (dB)	1.25	3.3	4.5
	SNR peak worst case (dB)	107	106	104
Complexity	# unitary capacitors	32	39	41
	# comparators	2	2	3
	# digital multipliers	1	2	3
	# digital adders	2	2	3
	# digital differenciators	2	2	5

In view of the results obtained, we can conclude that:

a) The finite gain as well as the capacitor mismatching degrade the ideal performance of cascade $\Sigma\Delta$ modulators, so that worst-case conditions should be considered during the design process.

b) In practice, since both architectures, 2-2 and 2-1-1, present similar behavior, a proper choice of coefficients considering circuit-level requirements, as well as technological aspects, will be crucial for the architecture selection.

c) For the cases covered, the classic option for the 2-2 modulator shows lower sensitivity to mismatching than the 2-2 and 2-1-1 architectures with optimized coefficients for maximum dynamic range and minimum OS in integrators.

6.3   IMPLEMENTATION OF THE 2-2 ARCHITECTURE

In this section the design process and test of a two-stage fourth-order 2-2 cascade ΣΔ modulator is described. The design cycle has been completed with the help of the CAD tools described in the previous chapters. Their use allows the advance from the modulator specifications up to the transistor sizes in only 30% of the total design time. The remaining 70% is necessary for completing the layout phase because it is carried out manually.

6.3.1   Determining the building block specifications

The architecture being designed is the 2-2 cascade of Fig. 6.2(a) with classic coefficient values:

$$g_1 = g_1' = g_2 = g_2' = g_3 = g_3'' = g_4 = g_4' = 0.5$$

$$g_3' = 0 \qquad d_0 = 1 \qquad d_1 = 4$$

(6.11)

The terminal specifications of the building blocks regarding the modulator specifications constitutes the first phase of the design process. Such modulator specifications are:

- Effective resolution: 17bit

- Digital output rate (DOR): 40kS/s (20-kHz signal bandwidth).

- Maximum input amplitude: 1V

These specifications serve as an input to the modulator sizing tool, SDOPT, as shown in the list of Fig. 6.8. The execution took 4-second CPU time on a Sparc10 work station. The results are shown in the SDOPT output file of Fig. 6.9. Note that the resolution obtained and that specified are in agreement. This fact, which is a consequence of the selection of a small value for the sampling capacitor (1pF), was deliberate in order to evaluate the accuracy of the models and optimization procedures of the methodology. In practice,

```
specify dsoc{
    type DSOC;
    bits 17;
    nyquistband 40e3;
    refvoltage 1.5;
}
variables {
    M   = 128;
    ADC = [3200, 5000, 500]-1.0;
    Io  = [10u, 50u, 5u]-1.0;
    Gm  = [200u, 1000u, 100u]-1.0;
    H   = [30m, 100m, 10m]1.0;
    R   = [1k, 5k, 500]1.0;
    NLC1 = 15; NLC2 = 0.0;
    ADC_NL1 = [1,10,0.1]1.0; ADC_NL2 = [20,50,5]1.0;
    Jitter = [2,5,0.1].5;
    C1 = 1p; Cp = 0.5p; Ci = 0.25p; Cl = 0.1p;
    G1 = 0.5; G1p = 0.5; G2 = 0.5; G3 = 0.5; G3pp = 0.5;
    nG1 = 4; mG1 = 8; nG2 = 4; mG2 = 8;
}
iterations {numitera 5000;}
spscan {
    print
    A*A/(2*(PQ+PTH+PHD_OP+PHD_C+PST+PSR_HD+PJITTER))
    PROGRAM XGRAPH db;
        A    range  1e-6 1.5 nstep 10 dec;}
```

Figure 6.8: SDOPT input file

the possibility of adjusting up to this extreme allows us to largely relax the specifications of the basic blocks, that are listed under the caption *Optimized Variables*. The error powers below are: distortion due to capacitor non-linearity (PHD_C), distortion due to the amplifier non-linear DC-gain (PHD_OP), jitter noise (PJITTER), quantization noise (PQ), slew-rate distortion (PHD_SR), incomplete settling noise (PST) and thermal noise (PTH). Among these, thermal noise is the dominant error source, which is typically the case for high resolutions (16 or more bit, see Section 5.8.1).

```
            DESIGN NAME -> dsoc
**Specifications**
Bits:          1.700000e+01
Nyquistband: 4.000000e+04
Refvoltage:   1.500000e+00
Order:        4
**Optimized variables**
ADC:         4.500000e+03      Gm:      4.000000e-04
ADC_NL1:     4.000000e+00      H:       5.000000e-02
ADC_NL2:     1.000000e+01      Io:      3.860000e-05
C1:          1.000000e-12      Jitter:  4.300000e-01
Ci:          2.500000e-13      M:       1.280000e+02
Cl:          1.000000e-13      NLC1:    1.500000e+01
Cp:          5.000000e-13      NLC2:    0.000000e+00
G1:          5.000000e-01      R:       5.000000e+02
G1p:         5.000000e-01      OS:      6.000000e+00
G2:          5.000000e-01      mG1:     8.000000e+00
G3:          5.000000e-01      mG2:     8.000000e+00
G3pp:        5.000000e-01      nG1:     4.000000e+00
                               nG2      4.000000e+00
**Resolution and Noise contributions**
Bits obtained          1.711893e+01
Equal to:              1.048159e+02   (dB) DR
PHD_C noise:           -1.105066e+02  (dB)
PHD_OP noise:          -1.141889e+02  (dB)
PJITTER noise:         -1.179482e+02  (dB)
PQ noise:              -1.284803e+02  (dB)
PSR_HD noise:          -1.298674e+02  (dB)
PST noise:             -1.418520e+02  (dB)
PTH noise:             -1.064686e+02  (dB)


**Iteration data**
NumItera:              5.000000e+03
CPU time (s):          4
```

Figure 6.9: SDOPT output file

The distortion caused by capacitor non-linearity, which is provided by the foundry (15ppm/V), is the second error source; although in practice its influence will be attenuated by using fully-differential circuits. Observe that the quantization noise, the only error contribution from the ideal point of view, is well below other non-ideal error sources. In this case, considering only the quantization noise would give an optimistic in more than 4bit estimate of the effective resolution. This facts corroborates the importance of evaluating the impact of the non-idealities for $\Sigma\Delta$ modulators with really demanding specifications.

In order to validate the results obtained with SDOPT, time-domain behavioral simulations have been performed using ASIDES. The input netlist is shown in Fig. 6.10. There, a model is associated with each basic block (integrators and comparators) whose parameters coincide with the specifications provided by SDOPT. Also included is the digital circuitry required to make the cancellation of the first-stage quantization, whose components (amplifiers/adders and delay blocks) have been supposed ideal.

Basically, two types of simulations are needed: on the one hand, the output spectrum for an input tone of variable frequency throughout the base band and an amplitude close to the maximum level. This simulation allows detection of the existence of harmonics and noise patterns as shown in Fig. 6.11(a).

On the other hand, in order to determine the dynamic range, the signal - (noise + distortion), (*TSNR*), after decimation filtering is simulated, while varying the input amplitude of a tone inside the base band. Fig. 6.11(b) corresponds to such a simulation. The good agreement between the analytical results and the simulations confirms the validity of sizing provided by SDOPT, especially those critical in this case, such as the dynamic requirements for the amplifiers and the number and value of the unitary capacitor.

```
#######################
# 2-2 Cascade Modulator #
#######################
# Primera etapa -> Second-order modulator
Vin inp ampl=(-110 4 2) freq=2.5k;
Comp1 out1 (oi2) real cm;
I1 oi1 (inp,out1*.5) real im;
I2 oi2 (oi1,out1*.5) real im;
# Segunda etapa -> Second-order modulator
Comp2 out2 (oi4) real cm;
I3 oi3 (oi2,out2*0.5) real im;
I4 oi4 (oi3,out2*0.5) real im;
# Digital cancellation of the quantization noise
Del3 14 (out1) full;
Del4 15 (14) full;
Add0 16 (15*1.0) ideal;
Add1 17 (out2*4.0 16*-1) ideal;
Del5 18 (17) full;
Ad2 19 (17 18*-1) ideal;
Del6 20 (19) full;
Ad3 21 (19 20*-1) ideal;
Ad4 22 (21 15) ideal;
.output snr(22) fft(22);
.clock freq=5.12X jitter=0.5ns;
.nsamp 65536;
.oversamp 128;
.options nofdt fullydiff;
# Models
.model im Integrator dcgain=4500 gm=400u cfb=2p cpa=0.5p
          cload=0.1p imax=38.6u cnl1=15u dcgnl1=4 dcgnl2=-10
          ron=500 osp=3 osn=-3;
.model cm Comparator vhigh=1.5 vlow=-1.5 hys=50m;
```

Figure 6.10: ASIDES netlist

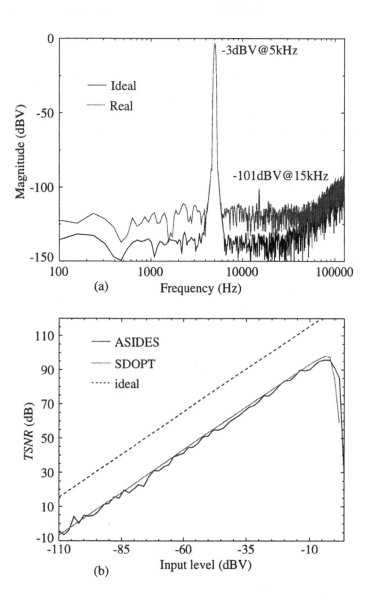

Figure 6.11: Behavioral simulation results: (a) Output spectrum for -3dBV, 5kHz input; (b) Signal-to-(noise+distortion) ratio after decimation as a function of the amplitude of an input tone at 2.5kHz.

6.3.2 Modulator schematic

Fig. 6.12 shows the fully-differential SC schematic of the modulator. It consists of the cascade connection of two second-order stages, each one with two identical integrators, with the exception of the fact that the first integrator (the one which receives the modulator input) includes chopper compensation of the offset and the flicker noise [Hsie81]. The whole circuit is controlled by two non-overlapped clock phases. During phase ϕ_1 (switches S_1 and S_3 ON) each integrator samples its input, which can either be the modulator input or the output of the previous integrator. Simultaneously, the comparators are strobbed because the output of the integrators does not change during this phase. During phase ϕ_2 (switches S_2 and S_4 ON) the integrators perform the subtraction operations and the results are accumulated in the integration capacitors. To avoid signal-dependent feedthrough, switches S_1 and S_2 are commuted with some delay with respect to S_3 and S_4 [Lee85]. In addition, phases controlling the chopper operation are included through switches S_d, S_c, S_{dd} and S_{cd}, that change the polarity of the amplifier during the non-over-lapping period of ϕ_1 and ϕ_2 according to the diagram of Fig. 6.13. The refer-

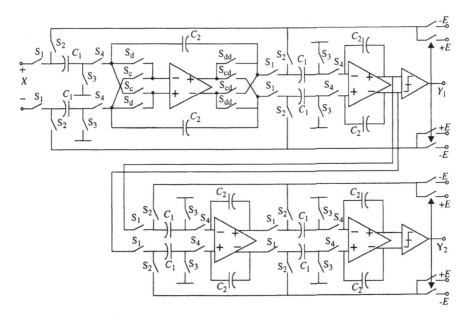

Figure 6.12: Fully-differential SC implementation of the analog part of the 2-2 modulator

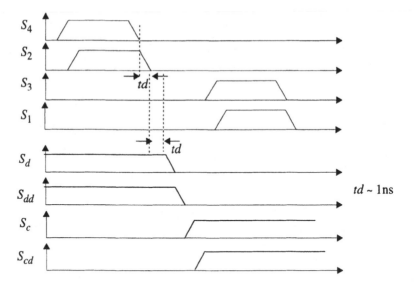

Figure 6.13: Clock phases

ence voltages are 1.5 V which equals $E = 0.75$ V in the fully-differential implementation. This schematic does not include the digital part of the modulator, see Fig. 6.2(a). It will be implemented by software, together with the decimation filter.

6.3.2.1 Amplifier

Fig. 6.14 shows the schematic of the amplifier used in the four integrators: a fully-differential folded-cascode OTA [Ribn84]. Since the DC-gain requirement is not so demanding (see SDOPT output file of Fig. 6.9), it is an interesting alternative because of its good operation speed / power consumption ratio. The common-mode was stabilized using a degenerated mirror-based structure whose linearity is good enough for our application and does not require extra power consumption [Duque93]. The schematic also includes the biasing stage formed by transistors M_{16} to M_{28}.

The amplifier was sized using FRIDGE to fulfil the specifications of the first column of Table 6.6. The results obtained after 35-min. CPU time are shown in Table 6.7. With this sizing, the electrical simulation of the amplifier using HSPICE provides the results of the second column of Table 6.6. It is worth emphasizing the good agreement between the simulation results and the specifications, trying to minimize the power consumption.

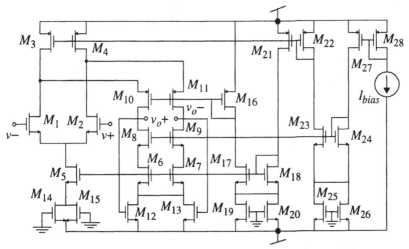

Figure 6.14: Fully-differential folded-cascode amplifier

Table 6.6: Specifications and simulation results for the amplifier

	Specs.	FRIDGE	Units
DC-gain	≥ 73	78.52	dB
gm	≥ 400	419	μA/V
GB(2pF)		34.88	MHz
MF(2pF)	≥ 60	66.28	o
I_o	≥38	40	μA
OS	≥ ±3	±3.2	V
Power consumption	minimize	1.95	mW

Table 6.7: Sizes for the amplifier of Fig. 6.14 (μm)

$M_{1,2}$	183.6/9.4	M_{16}	20/5
$M_{3,4}$	61.8/3	$M_{17,18}$	50/5
M_5	63.8/3	$M_{19,20}$	20/5
$M_{6,7}$	29/3	$M_{21,22}$	48/5
$M_{8,9}$	274.4/3	$M_{23,24}$	10/5
$M_{10,11}$	166.2/3	$M_{25,26}$	20/5
$M_{12,13}$	19.4/5	$M_{27,28}$	6.8/5
$M_{14,15}$	21.2/5	I_{bias}	40μA

6.3.2.2 Comparator

$\Sigma\Delta$ modulators show little sensitivity to the errors induced during the internal quantization. The position of the quantizer in the modulator loop causes these errors, the same as the quantization error, to be attenuated in the signal band. Such errors include offset, hysteresis, gain error and non-linearity. For the particular case of single-bit (two-level) quantization, the error possibilities are reduced to offset and hysteresis because it is not possible to define either gain or non-linearity.

With respect to the first, its effect is reduced to the appearance of a modulator offset with value equal to

$$V_{off, mod} = \frac{V_{off, comp}}{(A_V)^L} \tag{6.12}$$

where L stands for the order of the modulator loop and A_V is the amplifier open-loop low-frequency gain. For a second-order modulator the offset of the comparator is attenuated by A_V^2, which makes negligible the effect of typical values of the offset (up to some hundreds of millivolts). On the other hand, the hysteresis, as the indetermination of the output state for small input values (see Section 4.3.2.1) supposes an extra source of error, whose contribution to the total noise power is also attenuated in the signal band by the high DC-gain of the integrators. Thus, resolutions as low as 10% full scale do not degrade the modulator performance.

The little demanding resolution requirements, together with the necessary speed to provide the output in less than half of the sampling period, advise the use of a dynamic comparator based on a regenerative latch, like that of Fig. 6.15 [Yuka85]. This type of comparator typically have low resolution: hysteresis and offset around 50mV. However, that is sufficient and no pre-amplification stages are needed. The latch is activated at the end of phase ϕ_2. The output data are held in a NOR-gate SR flip-flop till the next phase ϕ_2. The latch was automatically sized using FRIDGE to fulfil the specifications in the first column of Table 6.8. Sizes in Table 6.9 were obtained after 15-min. CPU time. Again the electrical simulation results (second column of Table 6.8) agree with the specifications.

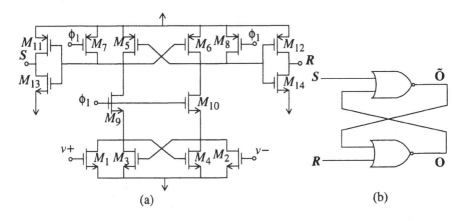

Figure 6.15: Comparator. (a) Regenerative latch. (b) SR flip-flop.

Table 6.8: Specifications and simulation results for the comparator

	Specs	FRIDGE	Units
TP_{HL}	< 24	8.0	ns
TP_{LH}	< 24	10.0	ns
Hysteresis	< 50	40	mV
Offset	–	77	mV

Table 6.9: Sizes for the latch of Fig. 6.15(a)

$M_{1,2}$	2/1.8	µm/µm	$M_{7,8}$	2.4/1.8	µm/µm
$M_{3,4}$	11/1.8	µm/µm	$M_{9,10}$	11.2/1.8	µm/µm
$M_{5,6}$	10.4/1.8	µm/µm	M_{11-14}	2/1.8	µm/µm

6.3.2.3 Clock phase generator

The generation of the clock phases is made on-chip by the circuit of Fig. 6.16. This provides two non-overlapped phases S_3 and S_4 together with two slightly delayed versions of them, S_1 and S_2, respectively. Such a small delay allows that the charge injection of the switches does not depend on the signal level [Lee85].

Chopper phases S_d (direct) and S_c (cross) are generated through edge-triggered D flip-flop with enable/disable input (DC). The two inverters pre-

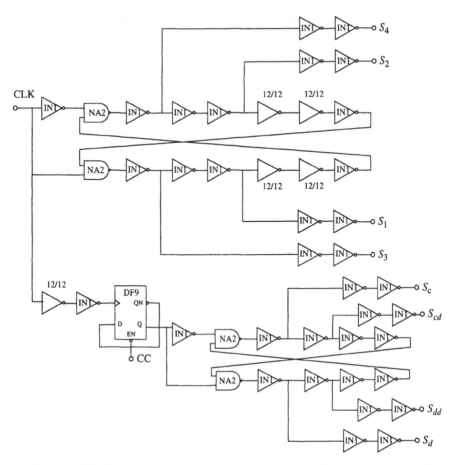

Figure 6.16: Clock phase generator with enable/dissable control of chopper operation

ceding the flip-flop make the chopper phases commute in the interval where all switches are OFF. These phases are non-overlapped (Fig. 6.13) to avoid a short circuit eventually being produced between both outputs of the amplifier. Also, two slightly delayed versions of the chopper phases S_{dd} and S_{cd} commute the output nodes of the integrator, which reduces the charge injection in the integration capacitor. Though not shown in the schematic for the sake of simplicity, the complementary phases of all the previous ones are also generated.

With the exception of the inverters labelled as 12/12 (12μm/12μm for both transistors PMOS and NMOS), all the cells of Fig. 6.16 are in the digital library provided by the foundry.

6.3.3 Design for testability

In order to increase the testability of the system, a test-chip was generated including the circuit of Fig. 6.17. It is basically a second-order modulator, the first of those which form the modulator of Fig. 6.12, together with 4 analog buffers connected through switches to the output of both integrators. The information that can be obtained with this arrangement is twofold:

- On the one hand, with switches S_{B1} to S_{B4} OFF we have a second-order ΣΔ modulator that can be independently tested.

- On the other, the access through buffers to the outputs of the integrators allows us to check on-chip the settling of the voltages in said points, including range problems, and to verify the operation of the comparator.

The analog buffer used for test purposes is that of Fig. 6.18 [Duque89]. This architecture, selected because it has very small input capacitance, was sized using FRIDGE (dimensions in Table 6.8). The simulation results in

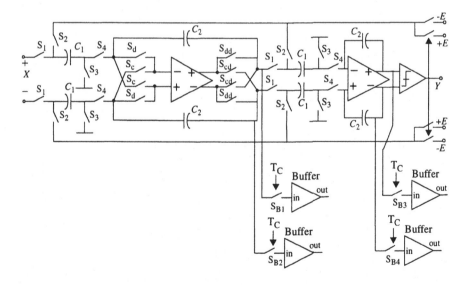

Figure 6.17: Circuit to increase testability: second-order modulator (first stage of the 2-2 modulador) and 4 analog buffers to scan the integrator outputs

Table 6.9, obtained with the extracted layout, show that, as expected, even though the frequency response is as required and the input capacitance very small, the positive output range is limited by the source follower (M_8). Even so, the use of two buffers allows an increase of the upper range of the differential output to 1.2V, which is enough to obtain the desired information.

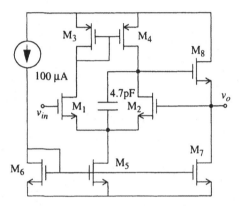

Figure 6.18: Analog buffer

Table 6.8: Sizes for the analog buffer (μm)

M1,2	48/2.2	M7	148.8/3
M3,4	167.2/2.2	M8	403.2/2.2
M5,6	20.8/2.2		

Table 6.9: Simulation results for the analog buffers

Specs.	FRIDGE	Units
f_{-3dB} [a]	34.35	MHz
C_{in} [b]	0.07	pF
Gain	-0.169	dB
Output swing	0.6 <-> -2.2	V
Power consumption	3.726	mW

a.with output load = 10pF.

b.at f_{-3dB}

6.3.4 Layout

Due to the coexistence in the same substrate of analog and digital signals, the layout of the prototypes is one of the critical phases in the design of ΣΔ modulators [Tsiv96]. Because of that, the layout, completed manually, consumed more than 70% of the total design time which, thanks to the use of the CAD methodology, was only three weeks.

The precautions adopted during the layout were the following:

• Use separate supply voltages for the digital and analog part.

• Avoid when possible the overlapping or proximity of analog and digital signals by choosing well differentiated routing channels.

• Place well and substrate contacts wherever there is proximity between analog and digital lines.

• Guard the analog cells by well/substrate contacts.

• Observe as much as possible the symmetry of the schematic.

• Partition in unitary transistors and use common-centroide techniques or similar for those transistors that have to be matched, making them drive the current in the same direction. Similar techniques should be used to lay-out the capacitors.

6.4   EXPERIMENTAL RESULTS

The measurement of high resolution converters requires using specific measurement set-up in an interference-free environment. Because of that, as a previous step to the presentation of the experimental results, the test set-up together with the equipment needed are described in detail. Afterwards, the results for the second-order modulator – the first stage of the fourth-order modulator, are presented. Finally, we include the corresponding results for the complete fourth-order modulator.

6.4.1   Description of the test set-up

Fig. 6.19 represents the set-up used to evaluate the modulator performance. A work-station simultaneously controls an HP82000 digital testing
unit and a high-quality programmable signal generator TM6003. The modulator, mounted on a specific printed board whose characteristics will be
described in the following paragraph, receives the clock signal and other control signals through the HP82000 unit. Jointly, it receives the input signal
from the programmable generator. The samples at the modulator output are
captured at the clock rate by the HP82000 and transferred to the work-station
for post-processing using MATLAB [Math91] or other signal processing
software. Controlling with the work-station the components of the set-up
allows us to automatically obtain the *SNR* against input level curves, which
are very useful to evaluate the modulator dynamic characteristics. Other controllable parameters are the clock frequency and the frequency and offset of
the modulator input signal.

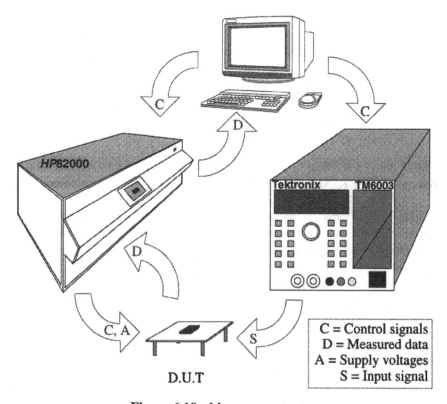

Figure 6.19: Measurement set-up

6.4.2 Printed board and environment features

The solutions adopted during the layout phase, in order to make compatible the coexistence of analog (sensitive to the noise) and digital (very noisy) circuitry in the same substrate, may be useless if no precautions are observed during testing. To this end, an environment as clean as possible should be provided for the chip. The external agents can interfere with the circuit under testing through three electromagnetic nature mechanisms [LaMa92]:

- *Direct coupling*, generated by the presence of undesired currents in the ground plane or in certain return lines. This type of coupling is responsible for most of switching noise (due to digital commutations) present in the analog lines.

- *Inductive coupling*, caused by magnetic fields that induce noisy current in the signal traces. The magnitude of these grows with the circuit loop area and with the proximity to the magnetic field source.

- *Capacitive coupling*, produced by parasitic capacitors which grow while the distance between lines and/or reference planes is reduced.

The main error mechanisms caused by the coupling above are:

- Presence of signals in the reference voltages with spectral components at half of the sampling frequency, due to coupling between clock lines and reference lines. Such signals are often used by the digital circuitry that are integrated with the modulator. Because of the multiplicative effect between the reference and modulator output voltage, when the input signal contains a small offset, spurious tones can be demodulated in the base band.

- Jitter noise. The effect of the clock period variation, that produces mainly white noise (Section 3.5.3), is favored by incorporation of auxiliary logic between the clock generator and the modulator.

In order to palliate such undesirable effects, the authors in [LaMa92] advise the use of certain techniques for the fabrication of specific printed boards, some of them have been considered in the design of Fig. 6.20 whose schematic is shown in Fig. 6.21.

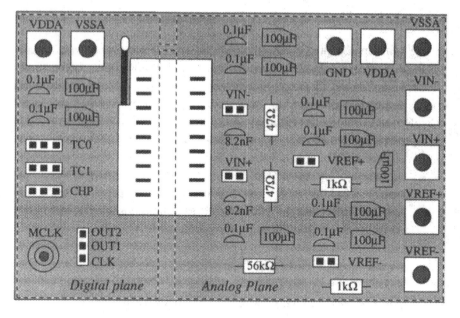

Figure 6.20: Printed board floor planning

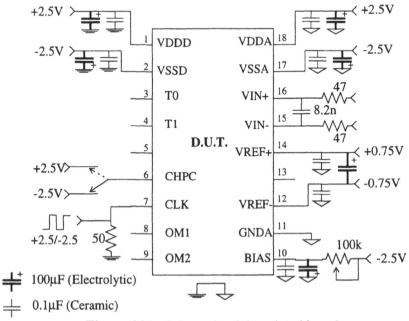

Figure 6.21: Schematic of the printed board

6.4.3 Experimental results

6.4.3.1 Second-order modulator

Fig. 6.22 shows a microphotograph of the second-order prototype in a CMOS 1.2μm technology. The circuit is biased with a voltage of 5V supplied by the HP82000 unit and its power consumption is of 5.5mW at 5-Mhz clock frequency (also supplied by the digital test equipment). The input signal is a tone with variable amplitude and frequency in the base band of the modulator, generated by the TM6003 unit. Such a signal is band limited through a first-order low-pass filter, just before entering the chip, in order to avoid aliasing (see pins VIN+ and VIN- in Fig. 6.21. The voltage of the reference pins is fixed to 0.75V , which taking into account that the circuitry is diferential, leads to modulator reference voltages of 1.5V .

Fig. 6.23 shows a three-dimensional representation of the modulator output spectrum with the amplitude of the input tone (at approximately 2.5kHz) as the second independent variable. To evaluate the power spectrum, 65,536 consecutive samples are collected from the modulator output using the HP82000 unit, while the amplitude and the frequency of the input are maintained constant. Once transferred to the work-station, each data series is processed using MATLAB, in this case through an FFT. Note that noise increases in the high-frequency zone (noise-shaping). Also, harmonic distortion appears for large amplitudes. This distortion (mainly of second order) is shown in greater detail in Fig. 6.24. The level of such a harmonic was very sensitive to changes in the configuration of the printed board, especially that of decoupling capacitors. Most probably it is not caused by the modulator itself but by deficiencies in measurement set-up.

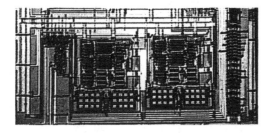

Figure 6.22: Microphotograph of the second-order modulator

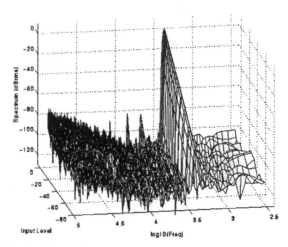

Figure 6.23: Output spectra for several signal levels

One of the possibilities of the modulator programmability consists of activating or de-activating the chopper operation in the first integrator. Fig. 6.24 shows that the offset is reduced by approximately 20dBrms when the chopper operation is activated.

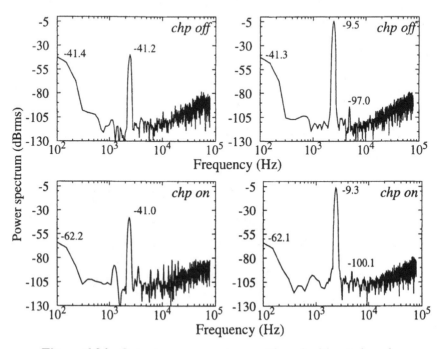

Figure 6.24: Output power spectrum with and without chopping

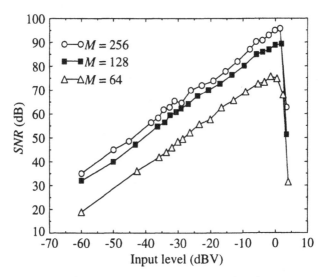

Figure 6.25: *SNR* vs. input level

Fig. 6.25 shows the signal-to-noise ratio (*SNR*) as a function of the input level. It was obtained using the TM6003 unit to generate a tone of frequency 5kHz (the clock frequency was fixed to 5MHz) with variable amplitude controlled from the work- station. For each amplitude, with the same procedure described above, 65536 samples were collected at the modulator output. These were stored in the work-station and processed with MATLAB. The processing consisted of low-pass filtering, using a third-order Sinc digital filter with cut-off frequency $f_S/(2M)$, where f_S represents the sampling frequency and M the oversampling ratio. The *SNR* curves show an effective resolution of 16bits (98dB dynamic range) for $M = 256$, that is 20kS/s digital output rate (*DOR*); 15bits (92dB) for $M = 128$, equivalent to 40kS/s *DOR*; and 13.1bits (81dB) for $M = 64$ or 80kS/s. Together with these data, Table 6.10 summarizes the modulator characteristics.

The access to the internal nodes through buffers allows us to check the correct settling of the voltages at the integrator outputs. For example, Fig. 6.26 shows the evolution of the first integrator output voltage (bottom), the modulator digital output (middle) and the clock (top).

Table 6.10:Summary of the second-order modulator performance

Oversampling ratio	256	128	64
DOR	20kS/s	40kS/s	80kS/s
Effective resolution (bit)	16.0	15.0	13.1
Dynamic range	98dB	92dB	81dB
SNR peak	95.7dB	89.3dB	76dB
TSNR peak	94.8dB	85.7dB	75.5dB
Maximum input level	1.0V		
Maximum sampling frequency	5.12 MHz		
Power average	5.5 mW		

6.4.3.2 Fourth-order modulator

Fig. 6.27 shows a microphotograph of the fourth-order prototype in a 1.2μm CMOS technology. It occupies an area of 0.94mm2 without pads and dissipates 10mW with a 5-V supply. The previously described measurement set-up and the printed board were also used for testing the fourth-order prototype. Fig. 6.28 shows the output spectrum of the fourth-order modulator for -9dBV, 4-kHz input sampled at 5.12MHz. Also, the output spectrum corre-

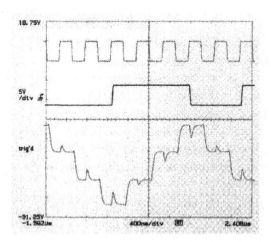

Figure 6.26: Integrator output monitorization

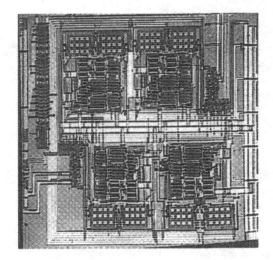

Figure 6.27: Microphotograph of the fourth-order modulator

sponding to first-stage output is shown. Observe the differences between the noise of the latter, second-order shaped, and that corresponding to the complete modulator, fourth-order shaped. However, as expected, the base band of the fourth-order modulator contains unshaped thermal noise because of the adjusted value of the sampling capacitor.

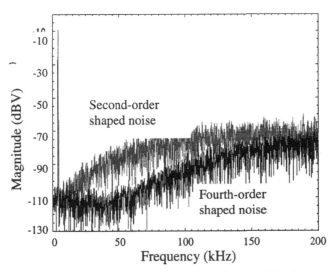

Figure 6.28: Output spectrum for -9dBV, 4-kHz input

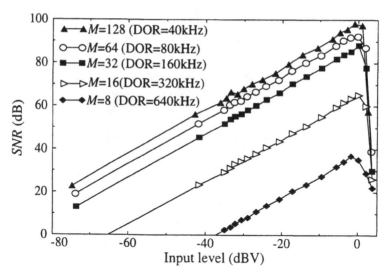

Figure 6.29: *SNR* vs. input level for several values of the oversampling ratio

Fig. 6.29 shows the SNR against input amplitude for five values of the oversampling ratio: 128 (nominal value), 64, 32, 16 and 8, which yields 40, 80, 160, 320 and 640kS/s digital output ratio, respectively. Observe that, for low oversampling ratios, the quantization noise dominates − the distance between consecutive curves reaches approximately the ideal value (obtained assuming that quantization is the only error source), which equals 27dB for a fourth-order modulator. For larger oversampling ratios, the dominant source of error is not quantization but unshaped white noise, therefore the separation between curves moves away from the theoretical value. The performance of the fourth-order modulator is summarized in Table 6.11.

SUMMARY

This chapter details the design process of a fourth order cascade $\Sigma\Delta$ modulator + with specifications of 17bit at 40kS/s.

To reach these specifications, first a comparative study is performed of fourth-order cascade architectures, from the ideal point of view as well as regarding the degree of sensibility to the non-idealities. Such a study concludes that a 2-2 cascade modulator formed by two second-order stages, apart from its greater simplicity, is the architecture less sensitive to mismatching in integrator weights.

Table 6.11:Summary of the fourth-order modulator performance

Oversampling ratio	128	64	32	16	8
DOR (kS/s)	40	80	160	320	640
Effective resolution (bit)	16.7	15.5	14.8	11	6.5
Dynamic range	102dB	95dB	91dB	68dB	41dB
SNR peak	98.2dB	92.5dB	88.2dB	65.1dB	36.8dB
TSNR peak	88dB	85dB	82dB	65dB	36.8dB
Maximum input amplitude	1V				
Maximum sampling rate	5.12 MHz				
Power (5-V supply)	10.0 mW				
Active area	0.94 mm^2				

The design of this modulator was carried out through the methodology proposed in Chapter 5, trying to adjust the specifications in the presence of error sources other than quantization, as for example thermal noise, with the secondary goal of testing the accuracy of the developed models.

Experimental results measured on a prototype fabricated in a 1.2μm CMOS technology show an effective resolution of 16.7bit at 40kS/s sampling at 5.12MHz clock frequency and with a power consumption of 10mW. The small deviation in the effective resolution is probably due to deficiencies of the test set-up also described in this chapter. Nevertheless, the performance of this modulator yields an *FOM* of 2.3pJ, which makes it state-of-the-art, as defined in Chapter 2.

Chapter 7

Integrated circuit design (II)

*A 13-bit 2.2MSample/s fourth-order cascade
multi-bit Sigma-Delta modulator*

7.1   INTRODUCTION

Most popular ΣΔ modulator architectures incorporate a simple comparator as a quantizer [Cand92], which ensures that the re-conversion to the analog plane is perfectly linear. The drawback of such gross conversion, which on the other hand provides very robust operation, is the large quantization noise, whose effect can be attenuated only by increasing the oversampling ratio.

However, the quantization noise decreases as quickly as the number of quantization levels increases [Paul87] with a factor inversely proportional to $(2^b - 1)^2$, where b stands for the number of bits of the internal conversion. Thus, the effective resolution increase of an ideal modulator with multi-bit quantization is $3.32\log(2^b - 1)$ bit. Furthermore, this gain in resolution is independent of the oversampling ratio, which is very interesting in high-frequency applications, where the speed of the analog circuitry is near the limits imposed by the technology. In addition, thanks to the larger number of quantization levels of the non-linear feedback, the architectures with multi-bit quantization present better stability and greater agreement with the analytical models [Teme93].

Unfortunately, in the case of multi-bit quantization, the D/A converter needed in the feedback loop is not linear per construction. Due to the position of this in the feedback loop (see Fig. 2.4a), its errors are injected, together with the signal, at the modulator input, which supposes that, like the signal itself, these errors are not attenuated in baseband. Because of that, small levels of D/A converter non-linearity cause an important degradation of the modulator performance.

Nevertheless, techniques exist for reducing the sensitivity of the A/D ΣΔ converters to the errors of the multi-bit quantization, which render these modulators an interesting alternative in many applications. These techniques can be classified as follows:

- Digital correction techniques, based on the cancellation in the digital domain of the errors induced during the in-loop D/A conversion [Cata89][Sarh93][Yang92].

- Dynamic matching techniques. The mismatching in the D/A converter elements (responsible for non-linearity) is corrected through a dynamic selection of these elements. The selection algorithms permit, in the simplest case, conversion of the harmonic distortion provoked by the non-linearity into white noise. In more elaborated applications, the spectral power density of this error can be shaped so that most of its power lies outside the signal band [Chen95][Bair96][Nys96].

- Techniques based on dual-quantization architectures. These architectures use single-bit and multi-bit quantization at different points of the topology. Their operation is based on the cancellation, in the digital domain, of the quantization noise associated with the coarsest quantization. In their simplest implementation, valid for any single-loop architecture, only the most significant bit is fed back, and then digitally combined with the remaining bits to form the multi-bit output code [Lesl90][Kiae93]. As with dynamic matching, this technique is intended to eliminate from the base band a part of the D/A conversion error power.

This chapter covers certain architectures belonging to this last category, which show the additional advantage of not requiring any extra analog circuitry. As will be seen, the multi-bit architectures generated are directly derived from the single-bit architectures, only by substituting a comparator (single-bit quantizer) with a multi-bit quantizer. Section 7.2 shows the pros and cons of the multi-bit quantization, and models are proposed for multi-bit A/D and D/A conversion errors, valid for automatic synthesis. Section 7.3 presents a dual-quantization cascade multi-bit $\Sigma\Delta$ modulator that improves those published up to now, concerning the sensitivity to the D/A conversion errors. Section 7.4 is devoted to a switched-capacitor implementation of such architecture in a $0.7\mu m$ CMOS technology. Experimental results are detailed in Section 7.5.

7.2   FUNDAMENTS OF MULTI-BIT QUANTIZATION

Fig. 7.1 shows the possible error mechanisms in a multi-bit A/D/A converter (assumed monotonic): offset error, gain error and non-linearity error [Jede89].

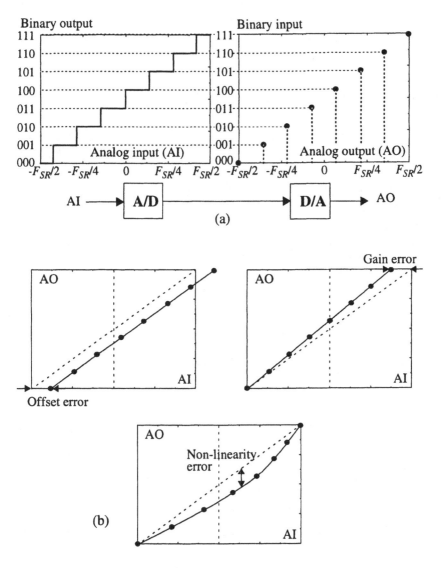

Figure 7.1:   (a) Series connection of an ideal 3-bit A/D and D/A converters. (b) Errors in the analog-analog transfer curve.

With illustrative purpose, Fig. 7.2 shows the impact of such errors in the signal-to-noise ratio (*SNR*) for a single-loop second-order modulator (Fig. 2.7) with 3-bit quantization. It is observed that neither the gain error nor the offset error constitute an important problem, because deviations as large as $0.5LSB_{3bit}$ do not produce *SNR* degradation. Exactly the opposite occurs with the non-linearity error: with no proper mechanism, even a small non-linearity value, such as $1mLSB_{3bit}$, provokes a 25% reduction in *SNR*. Thus, correction techniques are inevitably needed, whose development depends on the correct modeling of the errors above.

7.2.1   Modeling of the A/D and D/A conversion error

As stated, the error mechanisms of the A/D and D/A have a different impact on the performance of the multi-bit modulators. On the one hand, the A/D conversion errors, equivalent to those previously mentioned for the D/A converter, are treated like the quantization error: their power spectral density is displaced toward the high-frequency region, with which they play a secondary role. With respect to the D/A converter, whose errors are not filtered, the non-linearity is the main reason for the degradation of the modulator resolution. Other D/A conversion errors, like gain and offset errors, do not show any impact until they reach rather high values. That inclines the attention of

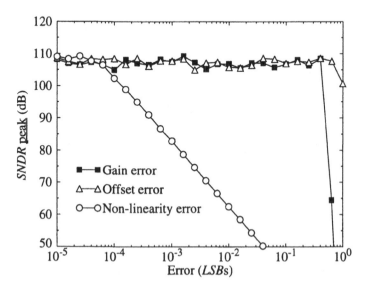

Figure 7.2: Signal-to-(noise+distortion) error vs. conversion errors

the designer toward the detailed analysis of the effect of the D/A converter non-linearity. However, the rest of the errors should be covered at least through behavioral simulation in order to verify the correctness of the analytical simplifications. In Chapter 4, ASIDES' behavioral models were obtained for multi-bit quantizers and D/A converters. Taking advantage of this knowledge, we will devote this section to obtaining analytical models to be included in the SDOPT's equation database.

Neglecting the offset and gain errors in expressions (4.27), and assuming that the D/A converter input is uniformly distributed in its full-scale range and that the number of the quantization levels is large, give us the following approximated expression for the non-linearity error power:

$$
\sigma_D^2 = \frac{1}{F_{SR}} \int_{-F_{SR}/2}^{F_{SR}/2} (w_a - x_a)^2 dx = \frac{1}{F_{SR}} \int_{-F_{SR}/2}^{F_{SR}/2} [\Phi(x_a) - x_a]^2 =
$$

$$
= \frac{1}{F_{SR}} \int_{-F_{SR}/2}^{F_{SR}/2} \left(-\varepsilon_0 x + \frac{\varepsilon_0}{A^2} x^3 \right)^2 dx \cong \frac{1}{2} (INL|_{LSB's})^2 q^2
$$

(7.1)

where F_{SR} stands for the input full-scale range, ε_0 and A were defined in (4.27), INL is the integral non-linearity and q is the separation between consecutive levels at the D/A converter output. As will be seen later, in spite of this approximation, which translates a non-linearity error as noise, and the previous linearization of the D/A converter operation, a good agreement is obtained between the analytical predictions and the simulated as well as measured results.

7.3   DUAL-QUANTIZATION CASCADE ARCHITECTURES

The cascade ΣΔ modulator architectures are natural candidates for the incorporation of multi-bit quantization in some of their quantizers. Recall that the principle of operation of the cascade modulators is the re-modulation, by means of a stage (low-order modulator), of the quantization noise generated in the previous stage/s. After digital cancellation of the first stage quantization noises, the quantization noise of the last stage is obtained with a noise-shaping function of an order equal to the number of integrators in all the stages.

7.3.1   Use of multi-bit quantization in cascade modulators

The idea, shown in Fig. 7.3, consists of including multi-bit quantization in the last stage of a cascade modulator, and single-bit quantization in the previous stage/s. Observe that the resulting modulator only differs from the cascade modulator of Section 3.2.1.1 in the number of bits of the last-stage quantizer; thus, the approximate expression obtained then for the Z-domain modulator output is still valid,

$$Y(z) = z^{-L_T}X(z) + d(1 - z^{-1})^{L_T}E_N(z) \qquad L_T = L_1 + L_2 + ... + L_N \qquad (7.2)$$

where $E_N(z)$ is the Z-transform of the quantization noise generated in a b-bit quantizer, which, under ideal conditions, provides the above-cited resolution improvement. As in single-bit quantization, d is a scalar, larger than unity, equal to the inverse of the product of the scaling factors (integrator weights) in the cascade (see Section 6.2.1.1).

The advantage of using multi-bit quantization only in the last stage, in comparison to the multi-bit single-loop modulators, is appraised taking into account the impact of the internal D/A conversion errors. In this architecture the errors caused by the D/A converter are not injected directly at the modulator input. Thus, they can be provided with a shaping function (like that of quantization error), so that most of its power is out of the signal band. This

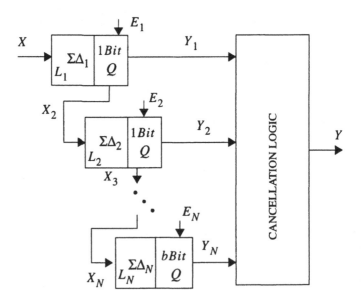

Figure 7.3: Generic cascade dual-quantization ΣΔ modulator

way, the Z-domain modulator output results in:

$$Y(z) = z^{-L_T}X(z) + d(1 - z^{-1})^{L_T}E_N(z) + d(1 - z^{-1})^{(L_T - L_N)}E_D(z) \quad (7.3)$$

where E_D is the Z-transform of the error produced in the last-stage D/A converter. Such an error presents a shaping function of an order $L_T - L_N$, which is enough to allow a certain degree of D/A converter non-linearity with no excessive degradation of the modulator performance.

7.3.2 Cascade modulators with multi-bit quantization

Fig. 7.4 and 6.5 show two cascade architectures with dual quantization reported previously. The first one, 2-1mb in Fig. 7.4, from Brandt and Wooley [Bran91b], consists of the multi-bit version of a two-stage third-

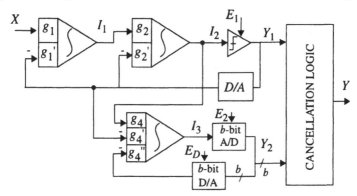

Figure 7.4: Third-order two-stage multi-bit modulator (2-1mb)

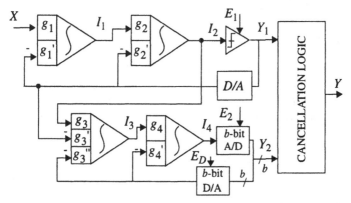

Figure 7.5: Fourth-order two-stage multi-bit modualtor (2-2mb)

order architecture [Will91]. The second one, 2- $2mb$, in Fig. 7.5, reported by Tan and Ericson [Tan93], incorporates multi-bit quantization in the second stage of a two-stage fourth-order modulator [Kare90][Bahe92]. The expressions equivalent to (7.3) for both architectures are

$$Y(z) \cong z^{-3}X(z) + d(1 - z^{-1})^3 E_2(z) - dz^{-1}(1 - z^{-1})^2 E_D(z) \quad \{2\text{-}1mb\} \tag{7.4}$$

$$Y(z) \cong z^{-4}X(z) + d(1 - z^{-1})^4 E_2(z) - d(1 - z^{-1})^2 E_D(z) \quad \{2\text{-}2mb\} \tag{7.5}$$

Note that, in either case, the errors due to the D/A conversion are, after digital cancellation, second-order shaped.

With the same philosophy, Fig. 7.6 shows an architecture that better exploits this technique. It is the multi-bit version of a three-stage fourth-order modulator 2-1-1, proposed in its single-bit version by Yin and Sansen [Yin94]. After digital cancellation, with the relationships of Table 7.1, the Z-domain modulator output results in:

$$Y(z) = z^{-4}X(z) + d_3(1 - z^{-1})^4 E_3(z) - d_3(1 - z^{-1})^3 E_D(z) \tag{7.6}$$

where the error due to the D/A conversion is third-order shaped. As a result, a larger insensitivity to such non-idealities is obtained. Integrating the noise contributions in expressions (7.4), (7.5) and (7.6) in the signal band, and neglecting provisionally other error sources, the following is obtained for the in-band noise power at the modulator output:

$$P_{2\text{-}1mb} = d^2\left(\sigma_Q^2 \frac{\pi^6}{7M^7} + \sigma_D^2 \frac{\pi^4}{5M^5}\right)$$

$$P_{2\text{-}2mb} = d^2\left(\sigma_Q^2 \frac{\pi^8}{9M^9} + \sigma_D^2 \frac{\pi^4}{5M^5}\right)$$

$$\sigma_Q^2 = \frac{q^2}{12}$$

$$P_{2\text{-}1\text{-}1mb} = d^2\left(\sigma_Q^2 \frac{\pi^8}{9M^9} + \sigma_D^2 \frac{\pi^6}{7M^7}\right)$$

$$\sigma_D^2 = \frac{1}{2}(INL|_{LSB's})^2 q^2 \tag{7.7}$$

where σ_Q^2 and σ_D^2 stand for the quantization error power and the D/A conversion error power, respectively, and M is the oversampling ratio. If we define the degree of sensitivity, S, to the D/A conversion non-idealities as the ratio between the contribution of such error to the total in-band error power and the ideal contribution (quantization noise), the following values are obtained for the three modulators:

$$S_{2\text{-}1mb} = \frac{7\sigma_D^2 M^2}{5\sigma_Q^2 \pi^2} \qquad S_{2\text{-}2mb} = \frac{9\sigma_D^2 M^4}{5\sigma_Q^2 \pi^4} \qquad S_{2\text{-}1\text{-}1mb} = \frac{9\sigma_D^2 M^2}{7\sigma_Q^2 \pi^2} \tag{7.8}$$

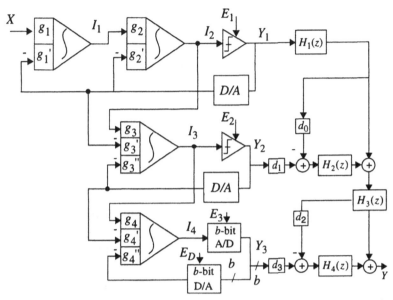

Figure 7.6: Fourth-order three-stage multi-bit modulator (2-1-1mb)

Table 7.1: Coefficient relationships and digital functions in Fig. 7.6

Digital	Digital/Analog	Analog
$H_1(z) = z^{-1}$	$d_0 = 1 - g_3'/(g_1 g_2 g_3)$	$g_1' = g_1$
$H_2(z) = (1 - z^{-1})^2$	$d_1 = g_3''/(g_1 g_2 g_3)$	$g_2' = 2g_1' g_2$
$H_3(z) = z^{-1}$	$d_2 = \left(1 - \dfrac{g_3'}{g_1 g_2 g_3}\right)\left(1 - \dfrac{g_4'}{g_3'' g_4}\right) \equiv 0$	$g_4' = g_3'' g_4$
$H_4(z) = (1 - z^{-1})^3$	$d_3 = g_4''/(g_1 g_2 g_3 g_4)$	

Thus, for a given value of M, σ_D and σ_Q, the 2-1-1mb architecture is $7M^2/(5\pi^2)$ times less sensitive to the non-idealities of the D/A converter than the 2-2mb architecture, while it approximately presents the same sensitivity as the 2-1mb architecture, although the absolute values of the in-band error powers for both differ in two orders of M. In those circumstances, the ratio between sensitivities shows a gain from 2 to 4bit in favor of the 2-1-1mb architecture, as compared to the 2-2mb, for M values from 16 to 64, and typical values of *INL* in the D/A converter. This is pointed out in Fig. 7.7 which shows the half-scale signal-to-(noise+distortion) ratio (*SNDR*) against the INL of a two-bit D/A converter for the previous architectures and $M = 16$. The analytical curves were obtained using the expressions in (7.7).

7.3.3   Influence of non-idealities

Chapter 3 described the special sensitivity of the cascade architectures to the integrator leakage and weight mismatching. The result of both non-idealities is an incomplete cancellation of the quantization noise generated in the first, or first and second stages, with the consequent *SNR* degradation. In a dual quantization architecture, the presence in the modulator output spectrum of the part of the quantization noise generated by single-bit quantization stages, imposes an upper limit to the number of bits to be used in the last-stage quantizer. Beyond that limit, which will depend on the value of such non-idealities, the benefits of a finer quantization in the last stage may be masked by the uncancelled portion of the quantization noise in previous stages. In particular, including the integrator leakage due to finite DC-gain and the mismatching in capacitor ratios, the in band error power for the architecture 2-1-1*mb* results in:

$$P_{2\text{-}1\text{-}1mb} = d^2\left(\sigma_Q^2 \frac{\pi^8}{9M^9} + \sigma_D^2\frac{\pi^6}{7M^7}\right) + \frac{\Delta^2}{12}\left(4\mu^2\frac{\pi^2}{3M^3} + \delta_A^2\frac{\pi^4}{5M^5}\right) \qquad (7.9)$$

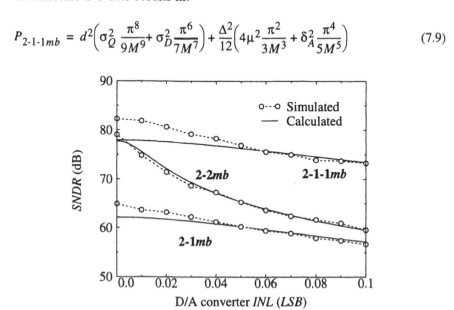

Figure 7.7:   *SNDR* against last-stage D/A converter *INL* for three cascade 2-bit architectures

where Δ represents the separation between comparator levels in the first two stages while μ and δ_A refer to leakage and mismatching, respectively[1].

For example, Fig. 7.8 shows the half-scale *SNR* obtained through behavioral simulation for a 2-1-1mb modulator, as a function of the number of bits of the last-stage quantizer. It corresponds to a worst-case simulation obtained through a Monte Carlo analysis, considering integrator weight standard deviation of 0.1%, DC-gain equal to 1000 and D/A converter *INL* equal to 1% of the full-scale range. The oversampling ratio is 16. Note that the curve saturate around 3 bit. Therefore, it does not make sense to use larger resolution in the last-stage quantizer.

7.4   IMPLEMENTATION OF THE 2-1-1*mb* ARCHITECTURE

The 2-1-1mb modulator of Fig. 7.6 is a novel architecture that presents lower sensitivity to the internal D/A conversion errors than other dual-quantization modulators reported up to now. This fact, that will allow use of very simple circuitry with consequent power saving, justifies by itself the electrical implementation of the 2-1-1*mb* modulator. Furthermore, the specifications:

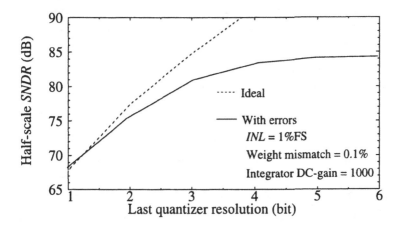

Figure 7.8:   *SNDR* degradation as a consequece of coefficient mismatching and integrator leakage

---

1. The effect of both non-idealities, integrator leakage and capacitor ratio mismatching, is analyzed in Section 3.2.1 and Section 3.2.2, respectively.

- Effective resolution: 12bit

- Digital output rate (DOR): 2.2MS/s (signal bandwidth: 1.1MHz)

- Maximum input level: ±1V

required to use the modulator in a ADSL system for copper-wire communications [ZCha95], will be more efficiently obtained including multi-bit quantization, because it allows reduction of the sampling frequency and, hence, the power consumption of the modulator.

7.4.1 Selection of integrator weights

The fulfillment of the relations in Table 7.1 is the basic premise for selecting the value of the analog coefficients (integrator weights in Fig. 7.6), in fact, whatever set of coefficients meeting such relations lead to the expression (7.6). Nevertheless for real implementation the following considerations must be taken into account:

- The level of the signal transferred from one stage to the next must be small enough not to overload the latter. These levels are equal to the reference voltages for a first-order modulator, and approximately 90% of the reference voltages for a second-order modulator.

- The output swing needed in the integrators, which depends on their weights as well as on the input level, must be physically achievable. In switched-capacitor implementations this limit is imposed by the supply voltages.

- The digital coefficient d_3, which amplifies the last-stage quantization error, must be as small as possible.

Additional considerations in order to simplify the implementation are:

- The digital coefficients should be 0, ±1 or multiple of 2.

- The gain of the last-stage quantizer, which, unlike that of the single-bit quantizers, is of importance in the multi-bit case, should not be larger than unity. As will be shown later, the larger the gain needed in the multi-bit quantizer, the more complex will be the A/D conversion.

As stated in Chapter 6, based on these criteria, the selection of the weight coefficients can be mapped into an optimization problem solvable by computational algorithms. The results are shown in Table 7.2. With these values, the integrator output swing requirement is reduced to only the reference voltages. In addition, the input of the last stage, given by $g_4 I_3(z) - g_4' Y_2(z)$, is just $E_2(z)$; that is, the input of the last stage does not contain any trace of the modulator input signal. Thus, the error due to a possible non-linearity of the last-stage D/A converter will not distort the modulated signal. This, absolutely true in the ideal case where coefficients are exactly those in Table 7.2, is still valid considering mismatching as stated in simulations of Fig. 7.9.

Table 7.2: Analog and digital coefficients in Fig. 7.6

g_1	0.25	g_3'	0.375	d_0	-2
g_1'	0.25	g_3''	0.25	d_1	2
g_2	0.5	g_4	4	d_2	0
g_2'	0.25	g_4'	1	d_3	2
g_3	1	g_4''	1		

7.4.2 Modulator sizing

Once the approximate error power expressions for all possible circuit non-idealities have been obtained, the synthesis at modulator level can be made automatically using SDOPT. The lists of Fig. 7.10 and Fig. 7.11 show the input and output files, respectively.

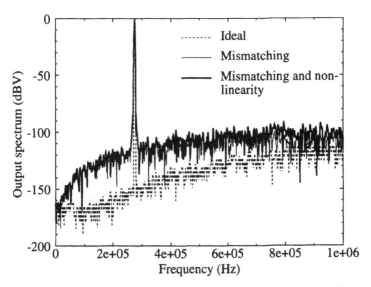

Figure 7.9: Output spectrum of a 3-bit 2-1-1mb modulator with $M = 16$ under ideal conditions and including coefficient mismatching ($\sigma = 0.5\%$) and non-linearity ($INL = 0.2LSB$)

```
specify md1{
  type SFFCMB;
  bits 12.5;
  nyquistband 2.2e6;
  refvoltage 2.0;
}
variables {
  M      = [16,32,8]-1.0;                    // oversampling ratio
  ADC    = [2000.0, 4000, 500]-1.0;          // Integrator DC-gain
  Io     = [600u, 3m, 500u]-1.0;             // maximum ouput current
  Gm     = [4e-03, 1e-2, 1e-3]-1.0;          // transconductance
  H      = [30m, 50e-3, 10e-3]1.0;           // comparator hystesis
  R      = [500, 1000, 100]1.0;              // switch ON resitance
  Cu     = [0.5p, 0.5p, 0.25p]1.0;           // unitary capacitor
  Cp     = 1.5p;                             // opamp input parasitic
  Cl     = 2.5p;                             // integrator output parasitic

  NLC1   = [25, 50.0, 1.0]1.0;               // Capacitor non-linearity (ppm/V)
  NLC2   = 0.0;
  ADC_NL1=[1,5.0,0.01]1.0;                   // First- r (%/V) and second-order(%/V²)
  ADC_NL2 = [5,20,0.5]1.0;                   // non-linearity of the amplifier DC-gain
  Jitter  = [0.1,.5,0.1]2.0;                 // clock period standard deviation (ns).
  N = 3;                                     // last stage quantizer resolution (bit)
  Inl = [0.2, 0.5, 0.1]1.0;                  // D/A converter non-linearity

  G1=0.25; G1p=0.25; G2=0.5; G3=1.0; G3pp=0.25; // g₁, g₁', g₂, g₃, g₃', g₃"
  G4=4; G4pp=1; // g₄, g₄"

  nG1 = 1;mG1 = 4; nG1p = 1; nG2 = 1;mG2 = 2; // # of unitary capacitor that forms the
  nG3 = 4; mG3 = 4; nG3pp = 1;                // numerator (n) and denominator (m)
  nG4 = 4; nG4pp = 1;                         // for each analog coefficient.

  A = 1;                                     // input level for distortion calculation
}
iterations {
  numitera 4000;
}
spscan {
  print A*A/(2*(PQ+PST+PSR_HD+PTH+PJITTER+PHD_OP))
                                             // Signal-to-(noise+distortion) ratio
       PROGRAM    XGRAPH db;
       A    range  1e-5 2 nstep 10 dec;       // against input amplitude
}
```

Figure 7.10: SDOPT input file to automatically obtain the building block
           specifications as a function of the 2-1-1mb modulator specifi-
           cations

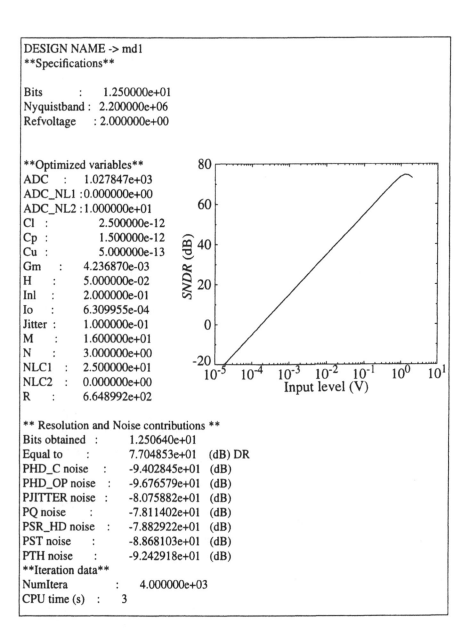

DESIGN NAME -> md1
Specifications

Bits   :   1.250000e+01
Nyquistband : 2.200000e+06
Refvoltage   : 2.000000e+00

Optimized variables
ADC   :   1.027847e+03
ADC_NL1 :0.000000e+00
ADC_NL2 :1.000000e+01
Cl   :   2.500000e-12
Cp :   1.500000e-12
Cu :   5.000000e-13
Gm   :   4.236870e-03
H   :   5.000000e-02
Inl   :   2.000000e-01
Io   :   6.309955e-04
Jitter :   1.000000e-01
M   :   1.600000e+01
N   :   3.000000e+00
NLC1   :   2.500000e+01
NLC2   :   0.000000e+00
R   :   6.648992e+02

** Resolution and Noise contributions **
Bits obtained :   1.250640e+01
Equal to   :   7.704853e+01   (dB) DR
PHD_C noise   :   -9.402845e+01   (dB)
PHD_OP noise :   -9.676579e+01   (dB)
PJITTER noise :   -8.075882e+01   (dB)
PQ noise   :   -7.811402e+01   (dB)
PSR_HD noise :   -7.882922e+01   (dB)
PST noise   :   -8.868103e+01   (dB)
PTH noise   :   -9.242918e+01   (dB)
Iteration data
NumItera   :   4.000000e+03
CPU time (s) :   3

Figure 7.11: SDOPT results: ouptut file and plot of theoretical signal-to-(noise+distortion) ratio

In the input file, the type of modulator is first specified (SFFCMB = 2-1-1mb). The modulator high-level specifications are included in the next three lines. The following block is dedicated to the independent variables of the synthesis process which coincide with the terminal specifications of the building blocks and other architecture parameters. The iteration count is fixed at 4,000 which is large enough to guarantee the convergence of the process. Finally, the $spscan$ block is intended to obtain graph information on the evolution of the signal-to-(noise+distortion) ratio ($TSNR$) as a function of the input level

SDOPT results are shown in Fig. 7.11. They yield an oversampling ratio of 16 with which places the clock frequency in 35.2MHz. The transconductance specified for the amplifiers implies a gain-bandwidth product (GB) around 135MHz for 5pF equivalent load. However, this load can vary from one to another phase and/or integrator. Because of that, it is interesting to preserve the transconductance specification instead that of GB. In addition to the resulting terminal specifications, information is given on the different noise and distortion contributions. In this case, apart from quantization, the main error sources are the slew-rate distortion and the jitter noise. Recall that according to the calculations of Section 3.5.3, this last one depends directly on the signal frequency. That is the reason why in this application 100ps. of standard deviation in the clock period significantly contributes to the total error power.

However, thermal noise contribution is not relevant. Because of that, and in order to reduce the amplifier dynamic requirements, the value provided by SDOPT for the sampling capacitor is only 0.5pF. Other contributions such as incomplete settling error and amplifier non-linear open-loop gain and non-linear capacitor distortion are well below the limit imposed by the effective resolution of the modulator. Observe that, as a result of the optimization procedure, the effective resolution obtained (12.506bits) is very close to that specified. The plot in Fig. 7.11 is the answer to the $spscan$ block.

7.4.3   Switched-capacitor implementation

The 2-1-1mb architecture has been implemented in a 0.7μm CMOS single-poly double-metal technology (analog capacitors are made by poly over n+ diffusion). Fig. 7.12 shows the fully-differential SC schematic of the analog part of the 2-1-1mb modulator. The first stage is obtained through the connection of two (single-branch and 2-branch) SC integrators. A differential comparator based in a regenerative latch acts as a single-bit quantizer.

The feedback signal is produced using two AND gates that control the

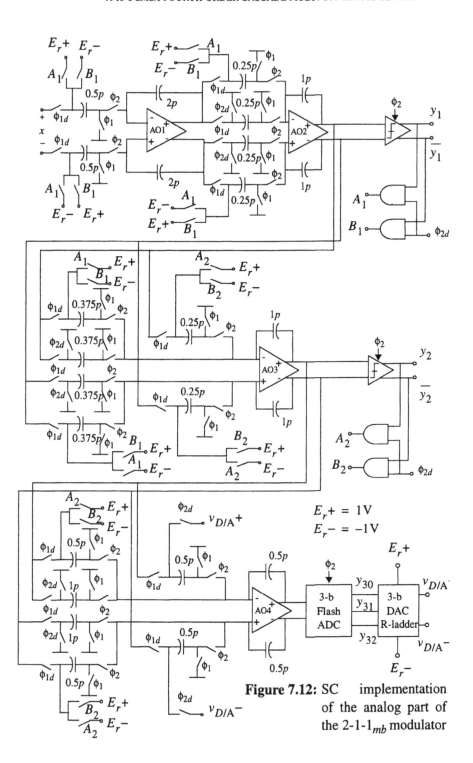

Figure 7.12: SC implementation of the analog part of the 2-1-1$_{mb}$ modulator

connection of the sampling capacitor of each integrator to $E_r + = 1V$ or $E_r - = -1V$ voltages, according to the sign of the comparator output; note that, because the circuitry is fully differential, it means that the reference voltage are $V_r = \pm 2V$. The second stage incorporates a 3-branch integrator to implement 3 different weights (g_3, g_3' and g_3''). The same architecture is used for the third-stage integrator, whose output drives a full-flash 3-bit A/D converter. The third-stage loop is closed through a 3-bit D/A converter implemented using a resistive ladder. In order to reduce the equivalent load of the second, third and fourth opamps, and hence their power consumption, the respective weights g_2, g_3 and g_4 have been distributed between two or three branches.

The operation of the modulator is controlled by two non-overlapped clock phases: during phase ϕ_1 the integrator input signals are sampled; during phase ϕ_2 the algebraic operations are performed and the results accumulated in the feedback capacitors of each integrator. The comparator and flash A/D converter are activated just at the end of phase ϕ_2 (controlling the strobe input with ϕ_2) to avoid any possible interference due to the transient response of the first- and second-stage integrator outputs in the beginning of phase ϕ_1. This timing guarantees a single delay per clock cycle. In order to attenuate the signal-dependent clock feedthrough, a set of slightly delayed versions of the two phases, ϕ_{1d} and ϕ_{2d}, are provided [Lee85].

7.4.4 Comparator

Because the hysteresis requirement for the comparator is not very demanding (see Fig. 7.11), while the resolution time should be a quarter of the sampling period (7ns), a simple regenerative latch is a good candidate to implement the single-bit quantizer. Fig. 7.13 shows such a comparator [Yuka85]. It is preferred to a latch with a pre-amplification stage, in order to minimize the power consumption. The latch was sized automatically using FRIDGE to fulfil the specifications in first column of Table 7.4. Sizes obtained after 15min. CPU are shown in Table 7.3. Simulation results are shown in the second column of Table 7.4.

Table 7.3: Comparator sizes in μm

Trans.	W/L	Trans.	W/L
$M_{1,2}$	3/4	$M_{7,8}$	2.2/0.7
$M_{3,4}$	3/4	$M_{9,10}$	2.2/0.7
$M_{5,6}$	8/3	$M_{11\text{-}14}$	2.2/0.7

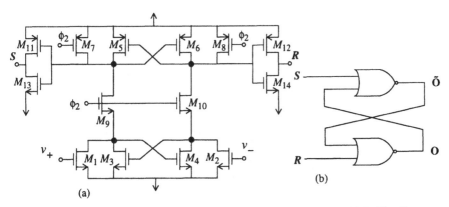

Figure 7.13: Comparator. (a) Regenerative latch. (b) NOR flip-flop.

Table 7.4: Comparator performance

	Specs.	Simulated	Units
Hysteresis	<50	< 10	mV
T_{PLH}	<7	6	ns
T_{PHL}	<7	6.5	ns
Power (average)	minimum	0.42	mW

7.4.5 A/D and D/A converters

The low sensitivity of the 2-1-1mb architecture to the imperfections in the D/A converter circuitry, and even more to those in the ADC, allows use of very simple topologies for both blocks, with the consequent power saving. In respect to the 3-bit A/D converter, a flash architecture [Lewi87] is suitable because data have to be coded into a small number of bits at the clock rate (35.2MHz). Fig. 7.14 shows three of the seven comparison stages of the converter. The classical architecture of these stages has been slightly modified to enable the comparison of differential signal, v_i, and reference, v_r, voltages, using two capacitors to compute the difference $(v_{i+} - v_{i-}) - (v_{r+} - v_{r-})$. At the end of phase ϕ_2, the comparator is activated to evaluate the sign of the difference. All the comparators are identical to that used in the first- and second-stage single-bit quantizer. In practice, without a pre-amplification stage, such a comparator may provide a resolution not much better than 50mV, which is enough to guarantee the *INL* requirement, taking into account that the value of the least significant bit (LSB) is 571mV. However, if the LSB decreases, as

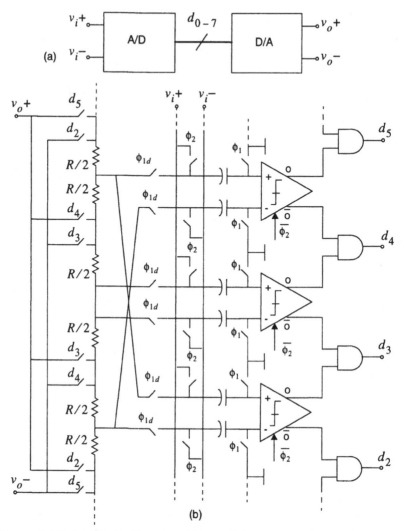

Figure 7.14: (a) Block diagram of the A/D/A system. (b) Partial view of its SC implementation.

would happen if the gain of the ADC had to be increased, the hysteresis inherent in the latched comparators might become a problem. In that case, more complex topologies with pre-amplification stages must be used, with a considerable increase in the power consumption. That is the reason why it is interesting to keep the ADC gain equal to unity. After the comparator array, eight AND gates generate a 1-of-8 code which is translated to binary by a ROM memory (not shown) whose outputs are buffered out of the chip. To implement the DAC, the 1-of-8 code is used to select through analog switches the voltages generated in a resistive ladder, also used to generate the reference voltages for the ADC. The resistances in the ladder are $R = 307.5\Omega$; this value is low enough to ensure that the settling error of the voltages in the ADC input capacitors (during phase ϕ_1) and fourth-integrator input capacitors (during phase ϕ_2) are not excessive. All switches are complementary with a size of 4.4μm/0.7μm for both NMOS and PMOS transistors.

7.4.6 Amplifiers

Fig. 7.15(a) shows the schematic of the amplifier used in integrators; it is a fully-differential folded-cascode OTA (transistors M_1 to M_{11}) and a biasing stage (transistors M_{12} to M_{17}). The common-mode feedback net is based on the SC circuit of Fig. 7.15(b). This dynamic circuit is advantageous in terms of power consumption in respect to the static networks when the operating frequency is high. Table 7.5 shows an estimation of the equivalent load of each amplifier, which has been calculated as $C_{eq} = C_i + C_p + C_l[1 + (C_i + C_p)/C_o]$, where C_i and C_o are the sampling and feedback capacitor, respectively; and C_p and C_l are the parasitics at the amplifier input and output, respectively.

Note that C_{eq} during phase ϕ_2 is considerably larger for the last amplifier because during this phase its output drives the 3-bit A/D converter. A possible solution to this problem is to place buffers at the output of the last integrator [Bran91b]. However, they require extra power consumption and can limit the useful output range. Furthermore, in order to compensate for gain errors and non-linearity, equivalent buffers should be placed between the

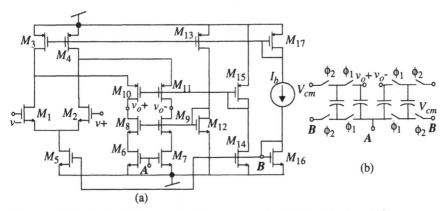

(a)

(b)

Figure 7.15: Fully-differential folded-cascode OTA: (a) Amplifier core. (b) SC common mode feedback circuit; all capacitor values are 0.2pF.

resistive ladder and the A/D comparison stages, which may suppose a considerable increase in power consumption. Another possibility is to design an amplifier for the first two integrators capable of driving from 4 to 6pF (fulfilling the specifications obtained from the modulator sizing) and a more powerful amplifier for the third and fourth integrator. We adopted the last solution because the availability of FRIDGE reduces significantly the design cycle, so that the design of two amplifiers, instead of just one, does not imply excessive extra time. Sizes obtained after 30min. CPU time are shown in Tables 7.6 and 7.7. The simulation results obtained from the extracted layouts of both amplifiers are shown in Table 7.8. Note that the supply current of the second amplifier is approximately twice that of the first. The phase margin of the first amplifier, 53degre, is large enough according to ASIDES behavioral simulations including a two-pole model for the amplifier. On the other hand, the in-band flicker noise (DC-1.1MHz) of the first opamp is 11mVrms (-99dBrms), which is well below the limit imposed by the resolution of the modulator. Thus, no low-frequency noise compensation mechanism is needed.

Table 7.6: Sizes of the amplifier in the first-stage integrators (μm)

Trans.	W/L	Trans.	W/L
$M_{1,2}$	492/1.2	M_{12}	2.3/1.2
$M_{3,4}$	137.7/1.2	M_{13}	13.8/1.2
M_5	331.2/1.2	M_{14}	18.2/1.2
$M_{6,7}$	182/1.2	M_{15}	3.1/1.2
$M_{8,9}$	19.6/1.2	M_{16}	66.1/1.2
$M_{10,11}$	142.8/1.2	M_{17}	26.2/1.2
		I_B	149μA

Table 7.7: Sizes of the amplifiers in second- and third-stage integrators (μm)

Trans.	W/L	Trans.	W/L
$M_{1,2}$	598.8/1.2	M_{12}	4.5/1.2
$M_{3,4}$	165/1.2	M_{13}	16.5/1.2
M_5	300/1.2	M_{14}	26.2/1.2
$M_{6,7}$	165/1.2	M_{15}	7.5/1.2
$M_{8,9}$	48/1.2	M_{16}	30/1.2
$M_{10,11}$	300/1.2	M_{17}	15.7/1.2
		I_B	143μA

Table 7.5: Estimation of the equivalent load per integrator and phase

Amplifier	phase ϕ_1	phase ϕ_2	Units
1	6.1	3.95	pF
2	5.8	3.25	pF
3	7.5	4.55	pF
4	3.25	11.575	pF

Table 7.8: Simulation results for the amplifiers

Specs.	1st & 2nd opamp	3rd & 4th opamp	Units
DC-gain	75.7	75.4	dB
Trasconductance	4.87	7.4	mA/V
GB	151(4pF)	93 (12pF)	MHz
PM	53 (4pF)	70.8 (12pF)	o
Input white noise	3.1	2.4	nV/√Hz
Input flicker noise (DC-1.1MHz)	11	-	mV$_{rms}$
Differential output swing	5.6	5.3	V
Maximum output current	0.7	1.35	mA
Supply current	1.82	3.4	mA

7.5   EXPERIMENTAL RESULTS

7.5.1   Internal A/D and D/A converters

The last-stage 3-bit A/D and D/A converters were integrated separately in a test chip to verify their operation. Fig. 7.16 shows the differential non-linearity (DNL) and integral non-linearity (INL) measured for several samples of the A/D and D/A converters, operating at 50MS/s. In both cases INL is less than 0.1LSB of 3bit, which leads to an equivalent resolution of 6bit. According to the results of SDOPT (INL < 0.2LSB), it is enough not to degrade the behavior of the complete modulator.

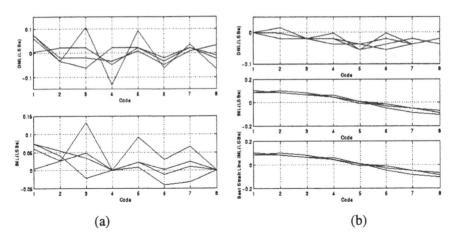

(a)   (b)

Figure 7.16: Measured linearity for the 3-bit (a) A/D converter, and (b) D/A converter.

7.5.2 2-1-1*mb* Modulator

Fig. 7.17 shows a microphotograph of the complete modulator, including the clock phase generator, fabricated in a 0.7μm CMOS technology. The prototype occupies 1.3mm² without the pads and dissipates 55mW operating at 5-V supply. The two-layer test board shown in Fig. 7.18 was used to characterize the modulator. It includes: separate analog and digital ground planes,

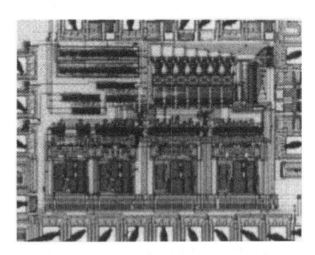

Figure 7.17: Microphotograph of the 2-1-1$_{mb}$ modulator in 0.7μm CMOS technology (area = 1.3mm²)

Figure 7.18: Printed board for testing the 2-1-1*mb* modulator

decoupling capacitors at the biasing and reference traces, first-order anti-alia-sing filter for the input with 2-MHz -3dB frequency and impedance coupling termination at the digital traces, in order to reduce the switching noise [LaMa92]. The performance of the modulator was evaluated using a high-quality programmable source to generate the input signal and a digital data acquisition unit to generate the clock signal and to acquire the bit streams of the first, second and third stages of the cascade. The same unit controlled the supply and reference voltages. After the acquisition, performed automatically by controlling the test set-up through proprietary C routines, data were transferred to a workstation to perform the digital postprocessing using MATLAB. The digital filtering was performed with a fifth-order *Sinc* filter, implemented by software.

Fig. 7.19 shows the modulator output spectrum after digital filtering obtained by processing 65,536 consecutive output samples at 35.2-MHz clock rate for a sinusoidal input of amplitude 0dBV and frequency 100kHz. Note that in Fig. 7.19(a), the noise in the base band (from DC to 1.1MHz) is nearly white (no noise-shaping is observed), most probably due to the presence of switching noise. In fact, as shown in Fig. 7.20, the in-band noise power strongly depended on the clock rate. We observed that it also varied with the supply voltage of the digital output buffers - the noisiest digital block. Thanks to the supply being provided separately for the digital and ana-

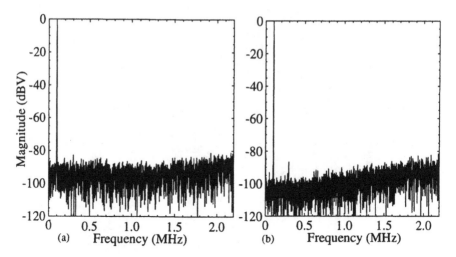

Figure 7.19: Measured output spectrums at 35.2-MHz clock-rate: (a) using 5-V supply and (b) 3-V supply for the digital output buffers.

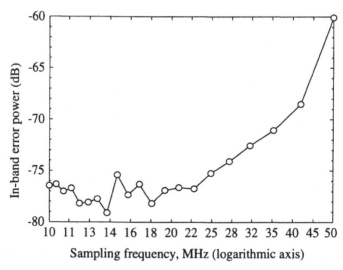

Figure 7.20: In-band error power against sampling frequency for $M = 16$ and digital supply = 5V

log part of the chip, we were able to significantly improve the modulator performance at the nominal sampling frequency by scaling down the supply voltage of the digital output buffers up to 3V. By doing so, the noise shaping was found at 35.2MHz, as shown in Fig. 7.19(b).

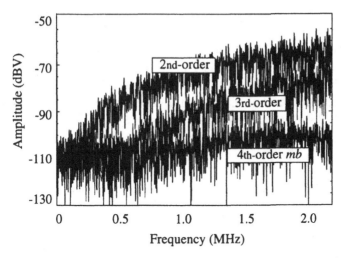

Figure 7.21: Comparison of different order noise-shaping

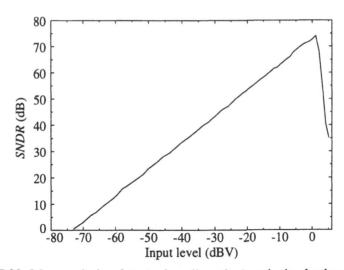

Figure 7.22: Measured signal-to-(noise+distortion) ratio in the base band (1.1MHz) as a function of the input amplitude

Fig. 7.21 compares the output spectrums of the complete modulator (labelled as 4th-order *mb*) and those corresponding to the first stage (2nd-order shaped) and to the combination of the first and second stages (3rd-order shaped).

Fig. 7.22 shows the signal-to-(noise+distortion) ratio (*SNDR*) in the base band (1.1MHz) as a function of the input level corresponding to the improved test set-up. Data were measured at 35.7-MHz clock rate with a sinusoidal input at 100kHz and amplitude varying from -73dBV to 6dBV. The maximum value of the *SNDR* was 74dB obtained for 1.1-dBV input amplitude. The in-band noise and distortion power yielded -76.5dB which, referred to the full-scale input (2V), is equivalent to 79.5-dB dynamic range or 12.9-bit effective resolution. The performance of the modulator is summarized in Table 7.9. Such a performance yields an *FOM* of 3.1pJ, which places this modulator between those with the lowest value of said figure reported until now (see Chapter 2).

Table 7.9: Summarized performance of the 2-1-1*mb* modulator

Sampling frequency	35.2	MHz
Digital output rate ($M = 16$)	2.2	MHz
Dynamic range	79.5	dB
Effective resolution	13	bit
SNDR peak	74	dB
THD peak	-90	dB
Reference voltages	2	V
Power consumption (5-V supply)	55	mW
Active area	1.3	mm^2

SUMMARY

This chapter has been devoted to exploring the application of $\Sigma\Delta$ modulation techniques for high-frequency A/D conversion. An approximation to the problem consists of the combination of high-order modulators with multi-bit quantization. In fact, the power in band of the quantization noise in a $\Sigma\Delta$ modulator is inversely proportional to $(2L + 1)M^{2L+1}$, M being the oversampling ratio and L the modulator order, and also inversely proportional to $(2^b - 1)^2$, where b is the number of bits of the internal quantization. Because of this, both techniques allow the reduction of the oversampling ratio required for obtaining a given resolution, which leads to an increase of the converter bandwidth, maintaining a constant sampling frequency.

The main drawback of the multi-bit quantization is the large sensitivity of the modulators to the non-linearity error of the D/A converter, necessary in the feedback loop. This problem is not present in single-bit quantization architectures, because in this case, the D/A conversion is linear per construction. Among the solutions proposed to palliate this effect there exists one which is especially interesting because it does not require extra circuitry for the correction of the non-linearity error. This solution consists of the use of cascade ΣΔ modulators incorporating multi-bit quantization in the last stage and single-bit quantization in the previous ones (*dual quantization*).

Based on this technique, we present a new ΣΔ modulator architecture which consists of a fourth-order 2-1-1 cascade modulator with multi-bit quantization in the third stage. For this modulator, the in-band error power of D/A conversion non-linearity is inversely proportional to M^7, instead of M^5, as in the previously reported cascade multi-bit ΣΔ modulators. Therefore, the 2-1-1*mb* modulator is less sensitive to the D/A conversion non-linearity than previous approaches.

The new architecture has been used to implement a 13bit, 2.2MS/s ΣΔ modulator. The design was carried out using the methodology presented in Chapter 5. Experimental results measured on a prototype fabricated in a 0.7μm CMOS technology, show an equivalent resolution of 12.9bits at 2.2MS/s, operating at 35.2MHz clock frequency and dissipating 55mW for 5-V supply. This performance yields an FOM equal to 3.1pJ, which is competitive with respect to the state-of-the-art converters summarized in Chapter 2.

REFERENCES

[Aaro56] M. Aaron: "The Use of Least Squares in System Design", *IRE Transactions on Circuit Theory*, Vol. 3, pp. 224-231, December 1956.

[Adam86] R. W. Adams: "Design and Implementation of an Audio 18-bit Analog-to-Digital Converter Using Oversampling Tecniques", *Journal of Audio Engeneering Society*, Vol. 34, pp. 153-166. March 1986.

[Adam91] R. W. Adams et al.: "Theory and Practical Implementation of a Fifth-Order Sigma-Delta A/D Converter", *Journal of Audio Engeneering Society*, Vol. 39, pp. 515-528, July 1991.

[Adam97a] R. W. Adams and R. Schreier: "Stability Theory for $\Sigma\Delta$ Modulators", Chapter 4 in the book *"Delta-Sigma Data Converters: Theory, Design and Simulation (S. R. Norswhorthy, R. Schreier and G. C. Temes, Editors)"*, pp. 141-163, IEEE Press, New York 1997.

[Adam97b] R. W. Adams: "The Design of High-Order Single-Bit $\Sigma\Delta$ ADCs", Chapter 5 in the book *"Delta-Sigma Data Converters: Theory, Design and Simulation (S. R. Norswhorthy, R. Schreier and G. C. Temes, Editors)"*, pp. 165-192, IEEE Press, New York 1997.

[Agra83] B. Agrawal and K. Shenoi: "Design Methodology for Sigma-Delta Modulators", *IEEE Transactions on Communications*, Vol. 31, pp. 360-369, March 1983.

[Ahm96] G.-C. Ahm et al.: "A 12-b, 10-MHz, 250-mW CMOS A/D Converter", *IEEE Journal of Solid-State Circuits*, Vol. 31, pp. 2030-2035, December 1996.

[Alle87] P. E. Allen and D. Holdberg: *"CMOS Analog Circuit Design"*. Holt, Rinehart Winston 1987.

[Anac91] Anacad Computer System: *ELDO: Electrical Circuit Simulator*, Ulm, Germany, 1991.

[Arda87] S. H. Ardalan and J. J. Paulos: "An Analysis of Non-linear Behavior in Delta-Sigma modulators". *IEEE Transactions on Circuits and Systems*, Vol. 34, pp. 593-603, June 1987.

[Au97] S. Au and B. H. Leung: "A 1.95-V, 0.34-mW, 12-b Sigma-Delta Modulator Stabilized by Local Feedback Loops", *IEEE Journal of Solid-State Circuits*, Vol. 32, March 1997.

[Bahe92] H. Baher and E. Afifi: "Novel Fourth-Order Sigma-Delta Convertor". *Electronics Letters*, Vol. 28, pp. 1437-1438, July 1992.

[Bair95] R.T. Baird and T.S. Fiez: "Improved $\Sigma\Delta$ DAC Linearity Using Data Weighted Averaging", *in Proc. of IEEE International Symposium on Circuits and Systems*, pp. 13-16, May 1995.

[Bair96] R. T. Baird and T. S. Fiez: "A Low Oversampling Ratio 14-b 500-kHz DS ADC with a Self-Calibrated Multibit DAC", *IEEE Journal of Solid-State Circuits*, Vol. 31, pp. 312-320, March 1996.

[Been92] G. F.M. Beenker, J. D. Conway, G. G. Schrooten and A. G.J. Slenter:

"Analog CAD for Consumer ICs", *in Proc. of the Workshop on Advances in Analog Circuit Design*, pp. 343-355, 1992.

[Benn48]   W. Bennett: "Spectra of Quantized Signals", *Bell Syst. Tech. J.*, Vol. 27, pp. 446-472, July 1948.

[Bert93]   A. Bertl, H. Leopold, P. O'leary and G. Winkler: "A Monolitic ADC for instrumentation Applications using Sigma-Delta Techniques", *in Proc. 11th European Conference on Circuit Theory and Design*, Vol. 2, pp. 1619-1624, 1993.

[Bose88a]   B. Boser, K. Karmann, H. Martin and B. Wooley: "Simulating and Testing Oversampled Analog-to-Digital Converters", *IEEE Transactions on Computer-Aided Design*, Vol. 7, pp. 668-674, June 1988.

[Bose88b]   B. E. Boser and B. A. Wooley: "The Design of Sigma-Delta Modulation Analog-to-Digital Converters". *IEEE Journal of Solid-State Circuits*, Vol. 23. pp. 1298-1308, December 1988.

[Bran91a]   B. Brandt, D. W. Wingard and B. A. Wooley: "Second-Order Sigma-Delta Modulation for Digital-Audio Signal Acquisition", *IEEE Journal of Solid-State Circuits*, Vol. 23. pp. 618-627, April 1991.

[Bran91b]   B. Brandt and B. A. Wooley: "A 50-MHz Multibit $\Sigma\Delta$ Modulator for 12-b 2-MHz A/D Conversion", *IEEE Journal of Solid-State Circuits*, Vol. 26, pp. 1746-1756, December 1991.

[Brau90]   G. Brauns et al.: "Table-Based Modeling of Delta-Sigma Modulators Using ZSIM", *IEEE Transactions on Computer-Aided Design*, Vol. 9, pp. 142-150, Feb. 1990.

[Bray90]   R. K. Brayton, A. Sangiovanni-Vincentelli and G.D. Hachtel: "Multilevel Logic Synthesis", *Proceedings of the IEEE*, Vol. 78, February 1990.

[Bren70]   P. R. Brent: *"Algorithms for Minimization without Derivatives"*, Englewood Cliffs, N. J., Prentice-Hall, 1970.

[Brid90]   N. Bridgett and C. P. Lewis: "Effect of initial conditions on limit cycle performance of second order sampled data sigma-delta modulator", *Electronics Letters*, Vol. 26, pp. 817-819, June 1990.

[Brod90]   R. W. Brodersen, et al.: "LAGER Tool Set", *User Manual*, UC Berkeley, 1990.

[Bult97]   K. Bult, A, Buchwald and J. Laskowski: "A 170mW 10b 50MSample/s CMOS ADC in 1mm^2" , *in Proc. of IEEE International Solid-State Circuit Conference*, pp. 136-137, 1997.

[Cand81]   J. C. Candy and O. J. Benjamin: "The structure of quantization noise from sigma-delta modulation", *IEEE Transactions on Communications*, Vol. 29, pp. 1316-1323, September 1981.

[Cand85]   J. C. Candy: "A Use of Double Integration in Sigma-Delta Modulation". *IEEE Transactions on Communications*, Vol. 33, pp. 249-258, March 1985.

[Cand92]   J. C. Candy and G. C. Temes, (Editors): *"Oversampling Delta-Sigma*

Converters". IEEE Press, 1992.

[Carl87]   L. R. Carley: "An Oversampling Analog-to-Digital Topology for High-Resolution Signal Adquisition Systems", *IEEE Transactions on Circuits and Systems*, Vol. 34, pp. 83-90, January 1987.

[Carl97]   L. R. Carley, R. Schreier and G. C. Temes: "Delta-Sigma ADCs with Multibit Internal Converters", Chapter 8 in the book *"Delta-Sigma Data Converters: Theory, Design and Simulation (S. R. Norswhorthy, R. Schreier and G. C. Temes, Editors)"*, pp. 244-281, IEEE Press, New York 1997.

[Cata89]   T. Cataltepe et al.: "Digitally Corrected Multi-bit $\Sigma\Delta$ Data Converters", *in Proc. of IEEE International Symposium on Circuits and Systems*, pp. 647-650, 1989.

[Chan94]   H. Chang et al.: "Top-Down, Constraint-Driven Design Methodology Based Generation of n-bit Interpolative Current Source D/A Converters", *in Proc. of IEEE Custom Integrated Circuits Conference*, pp. 369-372, 1994.

[Chen92]   F. Chen and B. Leung: "A Multi-Bit Σ-Δ DAC with Dynamic Element Matching Techniques", *in Proc. of IEEE Custom Integrated Circuits Conference*, pp. 16.2.1-16.2.4, May 1992.

[Chen95]   F. Chen and B. H. Leung: "A High resolution Multibit Sigma-Delta Modulator with Individual Level Averaging", *IEEE Journal of Solid-State Circuits*, Vol. 30, pp. 453-460, April 1995.

[Chou89]   W. Chou, P. Wong and R. Gray: "Multi-Stage Sigma-Delta Modulation", *IEEE Transactions on Information Therory*, Vol. 35, pp. 784-796, July 1989.

[Clin96]   D. W. Cline and P. R. Gray: "A Power Optimized 13-b 5 Msample/s Pipelined Analog-to-Digital Converter in 1.2µm CMOS", *IEEE Journal of Solid-State Circuits*, Vol. 31, pp. 294-303, March 1996.

[Cohn91]   M. Cohn, et al.: "KOAN/ANAGRAM II: New Tools for Device-Level Analog Placement and Routing", *IEEE Journal of Solid-State Circuits*, Vol. 26, March 1991.

[Conw92]   J. D. Conway and G.G. Schrooten: "An Automatic Layout Generator for Analog Circuits", *in Proc. of IEEE European Design Automation Conference*, pp. 513-519, 1992.

[Dedi94]   I. Dedic: "A Sixth-Order Triple-Loop $\Sigma\Delta$ CMOS ADC with 90dB SNR and 100kHz Bandwidth", *in Proc. of IEEE International Solid-State Circuits Conference*, pp. 188-189, 1994.

[Degr87]   M. G. R. Degrauwe et al.: "IDAC: An Interactive Design Tool for Analog CMOS Circuits", *IEEE Journal of Solid-State Circuits*, Vol. 22, pp. 1106-1114, December 1987.

[Dias91a]   V. F. Dias: "A Design Environment for Switched-Capacitor Noise-Shaping A/D Converters", *Ph. D. Dissertation*, Universita' Degli Studi di Pavia, 1991.

[Dias91b] V. F. Dias, V. Liberali and F. Maloberti: "TOSCA: a User-Friendly Behavioral Simulator for Oversampling A/D Converters". *in Proc. of IEEE International Symposium on Circuits and Systems*, pp. 2677-2680, 1991.

[Dias92a] V. F. Dias et al.: "Design Tools for Oversampling Data Converters: Needs and Solutions". *Microelectronics Journal*, Vol. 23, pp. 641-650, 1992.

[Dias92b] V. F. Dias, G. Palmisano, P. O'Leary and F. Maloberti: "Fundamental Limitations of Switched-Capacitor Sigma-Delta Modulators", *IEE Proceedings-G*, Vol. 139, pp. 27-32, February 1992.

[Duque89] J. F. Duque-Carrillo and R. Pérez-Aloe: "High-Bandwidth CMOS Test Buffer with Very Small Input Capacitance", *Electronics Letters*, Vol. 26, pp. 540-543, 1989.

[Duque93] J. F. Duque-Carrillo: "Control of the Common-Mode Component in CMOS Continuous-Time Fully Differential Signal Processing", *Analog Circuits and Signal Processing*, Vol. 4, pp. 131-140, 1993.

[El-Tu89] F. El-Turky and E. Perry: "BLADES: An Artificial Intelligence Approach to Analog Circuits Design". *IEEE Transactions on Computer-Aided Design*, Vol. 8, pp. 680-691, June 1989.

[Feel91] O. Feely et al.: "The Effect of Integrator Leak in $\Sigma\Delta$ Modulation". *IEEE Transactions on Circuits and Systems*, Vol. 38, 1293-1305, November, 1991.

[Fuji97] I. Fujimori et al.: "A 5-V Single-Chip Delta-Sigma Audio A/D Converter with 111dB Dynamic Range", *IEEE Journal of Solid-State Circuits*, Vol. 32, pp. 329-336, March 1997.

[Garr88] D. Garrod, R.A. Rutenbar and L.R. Carley: "Automatic Layout of Custom Analog Cells: ANAGRAM", *in Proc. of IEEE International Conference on Computer-Aided Design*, 1988.

[Geig82] R. Geiger and E. Sánchez-Sinencio: "Operational Amplifier Gain-Bandwdth Product Effects on the Performance of Switched-Capacitors Networks", *IEEE Transactions on Circuits and Systems*, Vol. 29, pp. 96-106, February 1982.

[Giel89] G. Gielen, H. Walsharts and W. Sansen: "ISAAC: A Symbolic Simulator for Analog Integrated Circuits", *IEEE Journal of Solid-State Circuits*, Vol. 24, pp. 1587-1597, December 1989.

[Giel90] G. Gielen, H. Walsharts and W. Sansen: "Analog Circuits Design Optimization Based on Symbolic Simulation and Simulated Annealing", *IEEE Journal of Solid-State Circuits*, Vol. 25, pp. 707-713, June 1990.

[Giel91] G. Gielen and W. Sansen: "*Symbolic Analysis for Automated Design of Analog Integrated CircuitsAnalog Circuits*", Kluwer, 1991.

[Giel92] G. Gielen and W. Sansen: "Open Analog Synthesis System Based on Declarative Models", *in Proc. of the Workshop on Advances in Analog Circuit Design*, pp. 397- 418, Scheveningen (The Netherlands), April 1992.

[Giel96]   G. Gielen and J. E. Franca: "CAD Tools for Data Converter Design: An Overview", *IEEE Transactions on Circuits and Systems-II*, Vol. 43, pp. 77-89, February 1996.

[Gobe83]   C. Gobet and A. Knob: "Noise Analysis of Switched Capacitor Networks", *IEEE Transactions on Circuits and Systems*, Vol. 30. January, 1983.

[Good96]   F. Goodenough: "Analog Technologies of all Varieties Dominate ISSCC", *Electronic Design*, Vol. 44, pp. 96-111, February 1996.

[Goods95]   M. Goodson, B. Zhang and R. Schreier: "Proving Stability of Delta-Sigma Modulators Using Invariant Sets", *in Proc. of IEEE International Symposium on Circuits and Systems*, pp. 663-636, 1995.

[Gray87]   P. R. Gray: "Analog IC's in the Submicron Era: Trends and Perspectives", *in Proc. of IEEE Electron Devices Meeting*, pp. 5-9, 1987.

[Gray90]   M. R. Gray: "Quantization Noise Spectra", *IEEE Transactions on Information Theory*, Vol. 36, pp. 1220-1244, November 1990.

[Gray93]   P. R. Gray and R. G. Meyer: "*Analysis and Design of Analog Integrated Circuits* (3er Edition)", Wiley, 1993.

[Greg81]   R. Gregorian: "High Resolution Switched-Capacitor D/A Converters", *Microelectronics Journal*, Vol. 12, pp. 10-13, 1981.

[Greg83]   R. Gregorian, K. W. Martin, G. C. Temes: "Switch-Capacitor Circuit Design", *Proceedings of the IEEE*, Vol. 71, pp. 941-964, August 1983.

[Gril96]   J. Grilo, E. Mac Robbie, R. Halim and G. Temes: "A 1.8V 94dB DR ΣΔ Modulator for Voice Applications", *in Proc. of IEEE International Solid-State Circuits Conference*, pp. 230-233, 1996.

[Hamm97]   C. Hammerschmied and Q. Huang: "A MOSFET-Only, 10b, 200kSample/s A/D Converter Capable of 12b Untrimmed Linearity", *in Proc. of IEEE International Solid-State Circuits Conference*, pp. 132-133, 1997.

[Harj89]   R. Harjani, R. Rutenbar and L.R. Carley: "OASYS: A Framework for Analog Circuits Synthesis", *IEEE Transactions on Computer-Aided Design*, Vol. 8, pp. 1247-1265, December 1989.

[Hash89]   M. Hashizume, H.Y. Kawai, K. Nii and T. Tameseda: "Design Automation System for Analog Circuits Based on Fuzzy Logic", *in Proc. of IEEE Custom Integrated Circuits Conference*, pp. 4.6.1-4.6.4, May 1989.

[Haus86]   M. W. Hauser and R. W. Brodersen: "Circuit and Technology Considerations for MOS Delta-Sigma A/D Converters", *in Proc. of IEEE International Symposium on Circuits and Systems*, May 1986.

[Hein92]   S. Hein et al.: "New Properties of Sigma-Delta Modulators with DC Inputs", *IEEE Transactions on Circuits and Systems*, Vol. 40, pp. 1312-1315, August 1992.

[Hsie81]   K. C. Hsieh, P. R. Gray, D. Senderowicz, and D. Messerschmitt: "A Low-Noise Chopper-Stabilized Differential Switched-Capacitor Filtering Technique", *IEEE Journal of Solid-State Circuits*. Vol. 16, pp. 708-715, December 1981.

[Hort91]   N. C. Horta, J. E. Franca and C. A. Leme: "FrameWork for Architecture Synthesis of Data Conversion Systems Employing Binary-Weighted Capacitor Arrays", *in Proc. of IEEE International Symposium on Circuits and Systems*, pp. 1789-1792, 1991.

[Inos62]   H. Inose, Y. Yasuda and J. Murakami: "A Telemetering System by Code Modulation-Δ–Σ Modulation", *IRE Transactions on Space Electronics and Telemetry*, Vol. 8, pp. 204-209. Septiembre, 1962.

[Ito94]   M. Ito et al.: "A 10 bit 20Ms/s 3 V Supply CMOS A/D Converter", *IEEE Journal of Solid-State Circuits*, Vol. 29, 1531-1536, December 1994.

[Jede89]   *"Terms, Definitions, and Letter Symbols for Analog-to-Digital and Digital-to-Analog Converters"*, JEDEC Standard No. 99, Addendum No. 1, July 1989

[John87]   D. S. Johnson, C. R. Aragon, L. A. McGeoch and C. Schevon: "Optimization by Simulated Annealing: an Experimental Evaluation", *Parts I and II, AT & T Bell Laboratories*, preprint, 1987.

[Jusu90]   G. Jusuf, P.R. Gray and A. Sangiovanni-Vincentelli: "CADICS - Cyclic Analog-To-Digital Converter Synthesis", *in Proc. of IEEE International Conference on Computer-Aided Design*, pp. 286-289, 1990.

[Jusu92]   G. Jusuf, P.R. Gray and A. Sangiovanni-Vincentelli: "A Performance-Driven Analog-to-Digital Converter Module Generator", *in Proc. of IEEE International Symposium on Circuits and Systems*, pp. 2160-2163, 1992.

[Kare90]   T. Karema, T. Ritoniemi and H. Tenhunen: "An Oversampled Sigma-Delta A/D Converter Circuit Using Two-Stage Fourth-order Modulator", *in Proc. of IEEE International Symposium on Circuits and Systems*, pp.3279-3282, 1990.

[Kaya88]   M. Kayal et al.: "SALIM: A Layout Generation Tool for Analog ICs", *in Proc. of IEEE Custom Integrated Circuits Conference*, pp. 7.5.1-7.5.4, 1988.

[Kenn88]   J. G. Kenney and L. R. Carley: "CLANS: A High-Level Synthesis Tool for High Resolution Data Converters", *in Proc. of IEEE International Symposium on Circuits and Systems*, pp.496-499, 1988

[Kenn93]   J. G. Kenney and L. R. Carley: "Design of Multibit Noise-Shaping Data Converters", *Analog Integrated Circuits and Signal Processing*, Vol. 3, pp. 99-112, 1993.

[Kert94]   D. A. Kerth et al.: "A 120dB Linear Switched-Capacitor Delta-Sigma Modulator", *in Proc. of IEEE International Solid-State Circuit Conference*, pp. 196-197, 1994.

[Kiae93]   S. Kiaei et al.: "Adaptive Digital Correction for Dual-Quantization $\Sigma\Delta$ Modulators", *in Proc. of IEEE International Symposium on Circuits and Systems*, pp. 1228-1230, May 1993.

[Kirk83]   S. Kirkpatrick, C.D. Gelatt and M.P. Vecchi: "Optimization by Simulated Annealing". *Science*, Vol. 220, pp. 671-680, May 1983.

[Kwak97]   S.-U. Kwak, B.-S. Song and K. Bacrania: "A 15b 5Msample/s Low-Spurious CMOS ADC", *in Proc. of IEEE International Solid-State Circuit Conference*, pp. 146-147, 1997.

[LaMa92]   J. L. LaMay and H. J. Bogard: "How to Obtain Maximum Practical Performance from State-of-the-Art Delta-Sigma Analog-to-Digital Converters", *IEEE Transactions on Instrumentation and Measurements*, Vol. 41, pp. 861-867, December 1992.

[Laar87]   P. J. M. Laarhoven and E.H.L. Aarts: "*Simulated Annealing: Theory and Applications*", Kluwer, 1987.

[Lee85]   K. Lee and R. G. Meyer: "Low Distortion Switched-Capacitor Filter Design Techniques", *IEEE Journal of Solid-State Circuits*, Vol. 20, pp. 1103-1113, December 1985.

[Lee87]   W. L. Lee and C. G. Sodini: "A Topology for Higher Order Interpolative Coders", *in Proc. of IEEE International Symposium on Circuits and Systems*, pp. 459-462, 1987.

[Leme93]   C. A. Leme: *Oversampled Interfaces for IC Sensors*, ETH Press, Zurich, 1993.

[Lesl90]   T .C. Leslie and B. Singh: "An Improved Sigma-Delta Modular Architecture", *in Proc. of IEEE International Symposium on Circuits and Systems*, pp. 372-375, May 1990.

[Leun88]   B. H. Leung et al.: "Area-efficient Multichannel Oversampled PCM Voice-Band Coder", *IEEE Journal of Solid-State Circuits*, Vol. 23, pp. 1351-1357, December 1988.

[Leun92]   B. H. Leung and S. Sutarja: "Multi-bit $\Sigma-\Delta$ A/D Converter Incorporating A Novel Class of Dynamic Element Matching Techniques", *IEEE Transactions on Circuit and Systems-II*, Vol. 39, pp. 35-51, January 1992.

[Leun97]   K. Y. Leung et al.: "A 5V, 118dB $\Delta\Sigma$ Analog-to-Digital Converter for Wideband Digital Audio", *in Proc. of IEEE International Solid-State Circuit Conference*, pp. 218-219, 1997.

[Lewi87]   S. Lewis and P. Gray: "A Pipelined 5-Msample/s 9-bit Analog-to-Digital Converter", *IEEE Journal of Solid-State Circuits*, Vol. 22, pp. 954-961, December 1987.

[Lim96]   S. Lim, S.-H. Lee and S.Y. Hwang: "A 12b 10MHz 250mW CMOS A/D Converter", *in Proc. of IEEE International Solid-State Circuit Conference*, pp. 316-317, 1996.

[Long88]   L. Longo and M. A. Copeland: "A 13-bit ISDN-band ADC using Two-Stage Third-Order Noise Shaping", *in Proc. of IEEE Custom Integrated Circuit Conference*, pp. 21.2.1-4, June 1988.

[Malo83]   F. Maloberti: "Switched-Capacitor Building Blocks for Analogue Signal Processing", *Electronics Letters*, Vol. 19, pp. 263-265, March 1983.

[Malo95] F. Maloberti, F. Francesconi, P. Malcovati and O. Nys: "Design Considerations on Low-Voltage Low-Power Data Converters", *IEEE Transactions on Circuits and Systems*-I, Vol. 42, pp. 853-863, November 1995.

[Man80] H. De Man et al.: "DIANA as a Mixed Mode Simulator for MOS LSI Sampled Data Circuits", *in Proc. of IEEE International Symposium on Circuits and Systems*, pp. 435-438, June 1980.

[Mant96] A. Mantyniemi, T. Rahkonen and A. Ruha: "A Low-Power 10-bit 300kS/s RSD Coded Pipeline A/D-Converter", *in Proc. of IEEE Workshop on Analog and Mixed IC Design*, pp. 79-82, September 1996.

[Mar95] M. F. Mar and R. W. Brodersen: "A Design System for On-Chip Oversampling A/D Interfaces", *IEEE Transactions on VLSI Systems*, Vol. 3, pp. 345-354, September 1995.

[Mart79] K. Martin and A. S. Sedra: "Strays-insensitive Switched-Capacitor Filters for a PCM Voice Codec.", *Electronics Letters*, Vol. 19, pp. 365-366, June 1979.

[Mart81] K. Martin and S. Sedra: "Effects of the Op Amp Finite Gain and Bandwidth on the Performance of Switched-Capacitor Filters", *IEEE Transactions on Circuits and Systems.* Vol. 28, pp. 822-829, August 1981.

[Math91] The MathWorks Inc.: "*MATLAB: User's Guide*", 1991.

[Mats87] Y. Matsuya et al.: "A 16-bit Oversampling A-to-D Conversion Technology Using Triple-Integration Noise Shaping", *IEEE Journal of Solid-State Circuits*, Vol. 22, pp. 921-929. December 1987.

[Mats94] Y. Matsuya and J. Yamada: "1V Power Supply, 384ks/s 10b A/D and D/A Converters with Swing-Suppression Noise Shaping", *in Proc. of IEEE International Solid-State Circuits Conference*, pp.192-193, 1994.

[Meta88] Meta Software Inc.: "*HSPICE User Manual*", 1988.

[Metr53] N. Metropolis et al.: "Equation of State Calculation by Fast Computing Machines", *Journal of Chem. Physics*, Vol. 21, pp. 1087-1092, 1953.

[Mino95] P. Minogue: "A 3 V GSM Codec", *IEEE Journal of Solid-State Circuits*, Vol. 30, pp. 1411-1420, December 1995.

[Mous94] S. M. Moussavi and B. H. Leung: "High-Order Single-Stage Single-Bit Oversampling A/D Converter Stabilized with Local Feedback Loops", *IEEE Transactions on Circuits and Systems*, Vol. 41, pp.19-25, January 1994.

[Naga85] K. Nagaraj, K. Sighal, T. R. Viswanathan and J. Vlach: "Reduction of Finite-Gain Effect on Switched-Capacitor Filters", *IEEE Transactions on Circuits and Systems*, Vol. 28, August 1985.

[Nage75] L. W. Nagel: "SPICE2: A Computer Program to Simulate Semiconductor Circuits", ERL-M520, University of California, Berkeley, 1975.

[Nors89] S. R. Norsworthy, I. G. Post and H. S. Fetterman: "A 14-bit 80-kHz Sigma-Delta A/D Converter: Modeling, Design and Performance Evaluation", *IEEE Journal of Solid-State Circuits*, Vol. 24, pp. 256-266, April 1989.

[Nors97a] S. R. Norsworthy, R. Schereier and G. C. Temes, (Editors): *"Delta-Sigma Data Converters: Theory, Design and Simulation"*, IEEE Press, New York 1997.

[Nors97b] S. R. Norsworthy: "Quantization Error and Dithering in $\Sigma\Delta$ Modulators", Chapter 3 in the book *"Delta-Sigma Data Converters: Theory, Design and Simulation (S. R. Norswhorthy, R. Schreier and G. C. Temes, Editors)"*, pp. 75-140, IEEE Press, New York 1997.

[Nye88] W. Nye et al.: "DELIGHT.SPICE: An Optimization-Based System for the Design of Integrated Circuits", *IEEE Transactions on Computer-Aided Design*, Vol. 7, pp. 501-519, April 1988.

[Nys93] O. Nys and E. Dijkstra: "Low Power Oversampled A/D Converters", *in Proc. 11th European Conference on Circuit Theory and Design*, Vol. 2, pp. 1595-1600, Davos, 1993.

[Nys96] O. Nys and R. Henderson: "A Monolithic 19bit 800Hz Low-Power Multi-bit Sigma Delta CMOS ADC using Data Weighted Averaging", *in Proc. of European Solid-State Circuits Conference*, pp. 252-255, 1996.

[Ocho94a] E. S. Ochotta: "Synthesis of High-Performance Analog Cells in ASTRX/OBLX", *Ph. D. dissertation*, 1994.

[Ocho94b] E. S. Ochotta, L.R. Carley and R.A. Rutenbar: "Analog Circuit Synthesis for Large, Realistic Cells: Designing a Pipelined A/D Converter with ASTRX/OBLX", *in Proc. of IEEE Custom Integrated Circuits Conference.*, pp. 365-368, 1994.

[Ocho96] E. S. Ochotta, R. A. Rutembar and L. R. Carley: "Synthesis of High-Performance Analog Circuits in ASTRX/OBLX", *IEEE Transactions on Computer-Aided Design of Integrated Circuits and Systems*, Vol. 15, pp. 273-294, March 1996.

[Okam93] T. Okamoto, Y. Maruyama and A. Yukawa: "A Stable High-Order Delta-Sigma Modulator with an FIR Spectrum Distributor", *IEEE Journal of Solid-State Circuits*, Vol. 28, pp. 730-735, July 1993.

[Opal96] A. Opal: "Sampled Data Simulation of Linear and Nonlinear Circuits", *IEEE Transactions on Computer-Aided Design of Integrated Circuits and Systems*, Vol. 15, pp. 295-306, March 1996.

[Op'T93] F. Op'T Eynde and W. Sansen: *Analog Interfaces for Digital Signal Processing Systems*, Kluwer 1993.

[Papo65] A. Papoulis: *"Probability, Randon Variables, and Stochastic Processes"*, McGraw-Hill, 1965.

[Paul87] J. J. Paulos, G. T. Brauns, M. B. Steer and S. H. Ardalan: "Improved Signal-to-Noise Ratio Using Tri-Level Delta-Sigma Modulation", *in Proc. of IEEE International Symposium on Circuits and Systems*, pp. 463-466, 1987.

[Pelu96] V. Peluso, M. Steyaert and W. Sansen: "A Switched Opamp 1.5V - 100μW $\Sigma\Delta$ Modulator with 12 bits Dynamic Range", *in Proc. European Solid-State Circuit Conference*, pp. 256-259, September 1996.

[Perr93]   D.L. Perry: "*VHDL, 2nd Edition*", MacGraw-Hill, 1993.

[Pill90]   L. T. Pillage and R.A. Rohrer: "Asymptotic Waveform Evaluation for Timing Analysis", *IEEE Transactions on Computer-Aided Design*, Vol. 9, pp. 352-366, April 1990.

[Plass78]   R. J. van de Plassche: "A Sigma-Delta Converter as and A/D Converter", *IEEE Transactions on Circuits and Systems*, Vol. 25, pp. 510-514, July 1978.

[Plass79]   R. J. van de Plassche: "A Monolitic 14-bit D/A converter", *IEEE Journal of Solid-State Circuits*, Vol. 14, pp. 552-556, June 1979.

[Rabi96]   S. Rabii and B. A. Wooley: "A 1.8V 5.4mW, Digital-Audio $\Sigma\Delta$ Modulator in 1.2μm CMOS", *in Proc. of IEEE International Solid-State Circuits Conference*, pp. 228-229, 1996.

[Ragh93]   V. Raghavan et al.: "AWE Inspired", *in Proc. of IEEE Custom Integrated Circuits Conference*, pp. 18.1.1-18.1.8, 1993.

[Rebe90]   M. Rebeschini et al.: "A 16-b 160 kHz CMOS A/D Converter using Sigma-Delta Modulation", *IEEE Journal of Solid-State Circuits*, Vol. 25, pp. 431-440, April 1990.

[Ribn84]   D. B. Ribner and M. A. Copeland: "Design Techniques for Cascode CMOS Op Amps with Improved PSRR and Common-Mode Input Range", *IEEE Journal of Solid-State Circuits*, Vol. 19, pp. 919-925, December 1984.

[Ribn91]   D. B. Ribner: "A Comparison of Modulator Networks for High-Order Oversampled $\Sigma\Delta$ Analog-to-Digital Converters", *IEEE Transactions on Circuits and Systems*, Vol. 38, pp. 145-159, February 1991.

[Rijm89]   J. Rijmenants, et al.: "ILAC: An Automated Layout Tool for Analog CMOS Circuits", *IEEE Jorunal of Solid-State Circuits*, Vol. 24, pp. 417-425, April 1989.

[Ritc77]   G. R. Ritchie: "Higher Order Interpolation Analog-to-Digital Converters", *Ph.D. Dissertation*, University of Pennsylvania, 1977.

[Rito88]   T. Ritoniemi, T. Karema, H. Tenhunen and M. Lindell: "Fully-differential CMOS Sigma-Delta Modulator for High-Performance Analog-to-Digital Conversion with 5 V Operating Voltage", *in Proc. of IEEE International Symposium on Circuits and Systems*, pp. 2321-2326, 1988.

[Rito90]   T. Ritoniemi et al.: "Design of Stable High Order 1-Bit Sigma-Delta Modulators", *in Proc. of IEEE International Symposium on Circuits and Systems*, pp. 3267-3270, 1990.

[Rohr67]   R.A. Rohrer: "Fully Automated Network Design by Digital Computer, Preliminary Considerations", *Proceedings of the IEEE*, Vol. 55, pp.1929-1939, December 1967.

[Rodr95]   A. Rodríguez-Vázquez and E. Sánchez-Sinencio, (Editors): "Special Issue on Low-Voltage and Low-Power Analog and Mixed-Signal Circuits and Systems", *IEEE Transactions on Circuits and Systems*, Vol.

42, November 1995.

[Rodr97] A. Rodríguez-Vázquez, F.V. Fernández and J.L. Huertas: "Introduction", Chapter 1 in the book *"Symbolic Analysis Techniques. Aplications to Analog Design Automation (F.V. Fernández, A. Rodríguez-Vázquez, G. Gielen and J.L. Huertas, Editors)"*, IEEE Press, 1997.

[Rued91] A. Rueda and J. L. Huertas: "Automated Analog Design", Chapter 18 in the book *"Analogue-Digital ASICs (R. S. Soin, F. Maloberti and J. Franca, Editors)"*, pp. 398-431, Peter Peregrinus Ltd., IEE, London, 1991.

[Rute89] R. A. Rutenbar: "Simulated Annealing Algorithms: An Overview" *IEEE Circuits and Devices Magazine*, Vol. 5, pp.19-26, January 1989.

[Saber87] *Saber User's Guide*, Analog Inc., Beaverton, OR. 1987.

[Sabi90] S. G. Sabiro, P. Senn and M.S. Tawfik: "HiFADiCC: a Prototype Framework of a Highly Flexible Analog To Digital Converter Silicon Compiler", *in Proc. of IEEE International Symposium on Circuits and Systems*, pp. 1114-1117, 1990.

[Sans87] W. Sansen et al.: "Transient Analysis of Charge Transfer in SC Filters: Gain Error and Distortion", *IEEE Journal of Solid-State Circuits*, Vol. 22, pp. 268-276, April 1987.

[Sarh93] M. Sarhang-Nejad, G.C. Temes: "A High-Resolution $\Sigma\Delta$ ADC with Digital Correction and Relaxed Amplifiers Requirements", *IEEE Journal of Solid-State Circuits*, Vol. 28, pp. 648-660, June 1993.

[Saue95] J. Sauerbrey and M. Mauthe: "12 Bit 1-mW $\Sigma\Delta$-Modulator for 2.4V Supply Voltage", *in Proc. of European Solid-State Circuit Conference*, pp. 138-141, 1995.

[Schr93] R. Schreier: "Desestabilizing Limit Cycles in Delta-Sigma Modulators", *in Proc. of IEEE International Symposium on Circuits and Systems*, Vol. 2, pp. 1369-1372, May 1993.

[Send97] D. Senderowicz et al.: "Low-Voltage Double-Sampled $\Sigma\Delta$ Converters", *in Proc. of IEEE International Solid-State Circuits Conference*, pp. 210-211, 1997.

[Shu95] T. Shu, B.-S. Song and K. Bacrania: "A 13-b 10-Msample/s ADC Digitally Calibrated with Oversampling Delta-Sigma Converter", *IEEE Journal of Solid-State Circuits*, Vol. 30, pp. 443-452, April 1995.

[Shyu84] J. B. Shyu, G. Temes and F. Krummenacher: "Random Error Effects in Matched MOS Capacitors and Current Sources", *IEEE Journal of Solid-State Circuits*, Vol. 19. pp. 948-955, December 1984.

[Song95] W.-C. Song et al.: "A 10-b 20-Msample/s Low-Power CMOS ADC", *IEEE Journal of Solid-State Circuits*, Vol. 30, pp. 514-521, May 1995.

[Spal96] J. Spalding and D. Dalton: "A 200Msample/s 6b Flash ADC in 0.6μm CMOS", *in Proc. of IEEE International Solid-State Circuit Conference*, pp. 320-321, 1996.

[Stee75] R. Stee'le: *"Delta Modulation Systems"*, Wiley, New York 1975.

[Suya90]   K. Suyama et al.: "Simulation of Mixed Switched-Capacitor/Digital Networks with Signal-Driven Switches", *IEEE Journal of Solid-State Circuits*, Vol. 25, pp. 1403-1413, December 1990.

[Taka91]   Y. Takasaki: *"Digital Transmission Design and Jitter Analysis"*, Artech House, Inc., 1991.

[Tan93]   N. Tan and S. Eriksson: "Fourth-Order Two-Stage Delta-Sigma Modulator Using Both 1 Bit and Multibit Quantizers", *Electronics Letters*, Vol. 29, pp. 937-938, May 1993.

[Tan95a]   N. Tan and S. Eriksson: "A Low-Voltage Switched-Current Delta-Sigma Modulator", *IEEE Journal of Solid-State Circuits*, Vol. 30, pp. 599-603, May 1995.

[Tan95b]   N. Tan: "A 1.2-V 0.8-mW SI $\Sigma\Delta$ A/D Converter in Standard Digital CMOS Process", *in Proc. of European Solid-State Circuit Conference*, pp. 150-153, 1995.

[Thom94]   C. D. Thompson and S. R. Bernadas: "A Digitally-Corrected 20b Delta-Sigma Modulator", *in Proc. of IEEE International Solid-State Circuit Conference*, pp. 194-195, 1994.

[Teme67]   G. C. Temes and D.A. Calahan: "Computer-Aided Network Optimization - The State-of-the-art", *Proceedings of the IEEE*, Vol. 55, pp. 1864, December 1967.

[Teme80]   C. G. Temes: "Finite Ampliffier Gain and Bandwidth Effects in Switched-Capacitor Filters", *IEEE Journal of Solid-State Circuits*, Vol. 15, pp. 358-360, June 1980.

[Teme93]   G. C. Temes and B. Leung: "$\Sigma\Delta$ Data Converter Architectures with Multibit Internal Quantizers", *in Proc. 11th European Conference on Circuit Theory and Design*, Vol. 2, pp. 1613-1618, Davos, 1993.

[Thom91]   D.E. Thomas: *"The Verilog Hardware Description Language"*, Kluwer, 1991.

[Tsiv96]   Y. Tsividis: *"Mixed Analog-Digital VLSI Devices and Technology"*, MacGraw-Hill, 1996.

[Touma94] C. Toumazou et al.: "Switched-Current Circuits and Systems", *in Proc. of IEEE International Symposium on Circuits and Systems - Tutorials*, pp. 460-486, May 1994.

[Uchi88]   K. Uchimura et al.: "Oversampling A-to-D and D-to-A Converters with Multistage Noise Shaping Modulators". *IEEE Transactions on Acoustics, Speech, and Signal Processing*, Vol. 36, pp. 1899-1905, December 1988.

[Vand84]   G. V. Vanderplaats: *"Numerical Optimization Techniques for Engineering Design: with Applications"*, McGraw-Hill, 1984.

[Vene96]   A. G. W. Venes and R. J. van de Plassche: "An 80MHz 80mW CMOS Folding A/D Converter with Distributed T/H Preprocessing", *in Proc. of IEEE International Solid-State Circuit Conference*, pp. 318-319, 1996.

[Vita92]   J. C. Vital and J. E. Franca: "Synthesis of High-Speed A/D Converter Architectures with Flexible Functional Simulation Capabilities", *in Proc. of IEEE International Symposium on Circuits and Systems*, pp. 2156-2159, 1992.

[Wang93]   H. Wang: "On the Stability of Third-Order Sigma-Delta Modulation", *in Proc. of IEEE International Symposium on Circuits and Systems*, pp. 1377-1380, 1993.

[Will91]   L. A. Williams, III and B. A. Wooley: "Third-Order Cascaded $\Sigma\Delta$ Modulators", *IEEE Transactions on Circuits and Systems*, Vol. 38, pp. 489-498, May 1991.

[Will94]   L. A. Williams, III , and B. A. Wooley: "A Third-Order Sigma-Delta Modulator with Extended Dynamic Range", *IEEE Journal of Solid-State Circuits*, Vol. 29, pp. 193-202, March 1994.

[Wolf90]   C. H. Wolff and L. Carley: "Simulation of $\Delta-\Sigma$ Modulators Using Behavioral Models", *in Proc. of IEEE International Symposium on Circuits and Systems*, pp. 376-379, 1990.

[Wu95]   C.-Y. Wu, C.-C. Chen and J.-J. Cho: "A CMOS Transistor-Only 8-b 4.5-Ms/s Pipelined Analog-to-Digital Converter Using Fully-Differential Current-Mode Circuit Techniques", *IEEE Journal of Solid-State Circuits*, Vol. 30, pp. 522-532, May 1995.

[Yang92]   Y. Yang, R. Schreier, G. C. Temes and S. Kiaei: "On-line Adaptive Digital Correction of Dual-Quantization Delta-Sigma Modulators", *Electronics Letters*, Vol. 28, pp. 1511-1513, July 1992.

[Yin92]   G. M. Yin, F. Op't Eynde, and W. Sansen: "A High-Speed CMOS Comparator with 8-Bit Resolution", *IEEE Journal of Solid-State Circuits*, Vol. 27, pp. 208-211, February 1992.

[Yin93a]   G. M. Yin, F. Stubbe and W. Sansen: "A 16-bit 320kHz CMOS A/D Converter using 2-Stage 3rd-Order $\Sigma\Delta$ Noise-Shaping", *IEEE Journal of Solid-State Circuits*, Vol. 28. pp. 640-647, June 1993.

[Yin93b]   G. M. Yin and W. Sansen: "A High-Frequency and High-Resolution Fourth-Order $\Sigma\Delta$ A/D Converter in BICMOS Technology", *in Proc. of European Solid-State Circuit Conference*, pp. 1-4, September 1993.

[Yin94]   G. Yin and W. Sansen: "A High-Frequency and High-Resolution Fourth-Order $\Sigma\Delta$ A/D Converter in Bi-CMOS Technology", *IEEE Journal of Solid-State Circuits*, Vol. 29, pp. 857-865, August 1994.

[Yots95]   M. Yotsuyanagi et al.: "A 2 V, 10 b, 20 Msample/s, Mixed-Mode Subranging CMOS A/D Converter", *IEEE Journal of Solid-State Circuits*, Vol. 30, pp. 1533-1537, December 1995.

[Youn90]   H. Young Koh, C.H. Sequin and P.R. Gray: "OPASYN: A Compiler for CMOS Operational Amplifier", *IEEE Transactions on Computer-Aided Design*, Vol. 9, pp. 113-125, February 1990.

[Yu96]   P. C. Yu and H.-S. Lee: "A 2.5V 12b 5Msample/s Pipelined CMOS ADC", *in Proc. of IEEE International Solid-State Circuits Conference*, pp. 314-315, 1996.

[Yuka85]   A. Yukawa: "A CMOS 8-bit High-Speed Converter IC", *IEEE Journal of Solid-State Circuits*, Vol. 20, pp. 775-779, June 1985.

[Yuka87]   A. Yukawa: "Constraints Analysis for Oversampling A-to-D Converter Structures on VLSI Implementation", *in Proc. of IEEE International Symposium on Circuits and Systems*, pp. 467-472, 1987.

[ZCha95]   Z-Y Chang, D. Macq, D. Haspeslagh, P. Spruyt and B. Goffart: "A CMOS Analog Front-End Circuit for an FDM-Based ADSL System", *IEEE Journal of Solid-State Circuits*, Vol. 30, pp. 1449-1456, April 1995.

[Zwan97]   E. J. van der Zwan: "A 2.3mW CMOS ΣΔ Modulator for Audio Applications", *in Proc. of IEEE International Solid-State Circuits Conference*, pp. 220-221, 1997.

Appendix (A)

Time-domain analysis of low-order ΣΔ modulators

A.1   INTRODUCTION

This appendix uses a technique similar to that in [Hein92] to determine an upper bound for the output signals of the integrators in a first- and second-order ΣΔ modulator, assuming DC inputs. However, the results obtained can be extended to the usual case in which the frequency of the input signal is small compared to the sampling frequency.

A.2   FIRST-ORDER MODULATOR

Let us consider the single-bit quantization first-order ΣΔ modulator of Fig. A.1. For a DC input with value X, the average of the modulator output signal, y, is given by

$$\bar{y} = \frac{v_1 - v_{-1}}{v_1 + v_{-1}} \cdot V_r = X \tag{A.1}$$

where V_r is the reference voltage (the D/A converter output levels are $\pm V_r$), and v_1 and v_{-1} are the number of positive and negative pulses, respectively, at the comparator output.

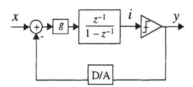

Figure A.1: First-order ΣΔ modulator

From (A.1)

$$\frac{v_1}{v_{-1}} = \frac{V_r + X}{V_r - X} \tag{A.2}$$

so that, in a pulse sequence of arbitrary length, the number of positive pulses increases as the input level approaches the reference voltage. In fact, in a first-order modulator, the number of consecutive positive pulses for input close to the reference value is given by the integer equal or greater than (A.2).

On the other hand, the finite difference equation describing the integrator output is

$$i_n = i_{n-1} + g(X - y_{n-1}V_r) \tag{A.3}$$

which can be re-written as

$$i_n = I_0 + \sum_{k=1}^{n} (i_k - i_{k-1}) = I_0 + g \sum_{k=1}^{n} (X - y_{k-1}V_r) \tag{A.4}$$

$$= I_0 + g\left(nX - V_r \sum_{k=0}^{n-1} y_k\right)$$

where $I_0 = i_{n=0}$. Assume that the first pulse of a sequence of consecutive positive pulse is obtained for $n = 0$. According to the previous considerations, the last pulse of the sequence is obtained for

$$n_l = \left\lfloor \frac{V_r + X}{V_r - X} \right\rfloor - 1 \tag{A.5}$$

where $\lfloor a \rfloor$ denotes the integer equal or greater than a. Thus, for $n = n_l + 1$ a negative pulse is obtained at the modulator output, which implies that $i_{n_l + 1} \le 0$. Applying (A.4) yields

$$i_{n_l + 1} \cong I_0 - g(V_r - X)\frac{V_r + X}{V_r - X} = I_0 - g(V_r + X) \le 0 \Rightarrow I_0 \le g(V_r + X) \tag{A.6}$$

where $\lfloor a \rfloor \cong a$; for $a \gg 1$ has been assumed; the closer the input to the reference level, the more exact such approximation. Repeating the previous calculations for negative inputs, we arrive at the following conclusion:

$$|I_0| \le g(V_r + |X|) \tag{A.7}$$

An analysis of expressions (A.4) and (A.7) points out that, for positive input close to the reference level, the integrator output evolves as a saw-teeth curve with the minimum slightly below the zero level and the maximum at $g(V_r + X)$. Alternatively, for negative inputs, the maxim of the modulator output slightly exceeds the zero level, while its minimum is at $-g(V_r - X)$. Thus, the worst case input (when $|X| \to V_r$) requires the following integrator output swing:

$$OS_1 = \pm 2g V_r \tag{A.8}$$

A.3   SECOND-ORDER MODULATOR

Consider the second-order ΣΔ modulator of Fig. A.2. In such an architecture with two in-loop integrations, obtaining an expression similar to (A.8) is more difficult because, though (A.1) and (A.2) are still valid, the length of the sequences of positive or negative pulses may be different from the integer equal or larger than (A.2), even for inputs close to the reference level. In fact, for a given input level, the modulator output sequences contain sub-sequences of variable length. In addition, the number of sub-sequences with different length strongly depends on the integrator initial conditions and on the input level [Brid90]. Fig. A.3 illustrates such a situation through a behavioral simulation for $X = 0.725V_r$. Because of such difficulty, we will use a fast-convergence iterative procedure in order to obtain the integrator output swing requirement.

Let us begin with the difference equation for the second-order modulator:

$$i_n = i_{n-1} + g_1(X - y_{n-1}V_r)$$
$$j_n = j_{n-1} + g_2(i_{n-1} - 2g_1 y_{n-1} V_r) \tag{A.9}$$

which can be re-written as

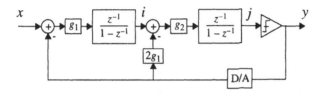

Figure A.2: Second-order ΣΔ modulator

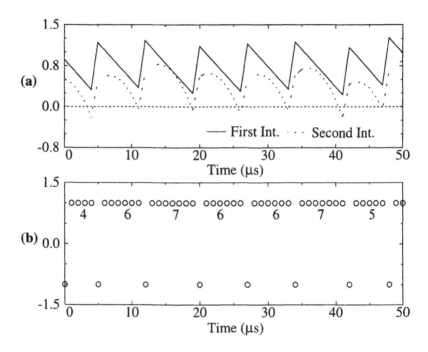

Figure A.3: (a) Output of both integrators. (b) Output sequence of the second-order modulator

$$i_n = I_0 + \sum_{k=1}^{n} (i_k - i_{k-1}) = I_0 + g_1\left(nX - V_r\sum_{k=0}^{n-1} y_k\right)$$

$$j_n = J_0 + \sum_{k=1}^{n} (j_k - j_{k-1}) = J_0 + g_2\left(\sum_{k=0}^{n-1} i_k - 2g_1 V_r\sum_{k=0}^{n-1} y_k\right)$$

(A.10)

Using the first equation in (A.10) yields

$$\sum_{k=0}^{n-1} i_k = nI_0 + g_1\left(X\sum_{k=0}^{n-1} k - V_r\sum_{k=0}^{n-1}\sum_{l=0}^{k-1} y_l\right) =$$

$$= nI_0 + \frac{n(n-1)}{2}g_1 X - g_1 V_r\sum_{k=0}^{n-2} (n-k-1)y_k \qquad n \geq 2$$

(A.11)

by with, the second equation in (A.10) results in:

$$j_n = J_0 + g_2 n I_0 + g_1 g_2 \frac{n(n-1)}{2} X - 2 g_1 g_2 V_r y_{n-1} \qquad (A.12)$$

$$+ g_1 g_2 V_r \sum_{k=0}^{n-2} (n-k+1) y_k$$

Let X be a positive input close to V_r. Starting from a positive initial condition in both integrators, positive pulses are generated until the second integrator output becomes negative. Then, a negative pulse is produced at the modulator output. This happens for the first value of n fulfilling

$$J_0 + g_2 n I_0 + g_1 g_2 \left(\frac{n(n-1)}{2} X - \frac{n(n+3)}{2} V_r \right) \le 0 \qquad (A.13)$$

which is met by making $y_k = 1$, $k = 0, 1, \ldots, n-1$ in (A.12). Such a value of n is

$$n_0 = \lfloor r_0 \rfloor \qquad (A.14)$$

where r_0 is the value (integer or non-integer) of n for which the equality in (A.13) is reached. The negative pulse at $n = n_0$ increases the output of both integrators that become positive again, starting another sequence of positive pulses. The duration of such a sequence, which may differ from the previous one, depends on the integrator output values at $n = n_0 + 1$ – the new initial conditions in (A.12). These values are known:

$$I_0 = I_{n_0} + g_1(X + V_r)$$
$$J_0 = J_{n_0} + g_2(I_{n_0} + 2g_1 V_r) \qquad (A.15)$$

Fig. A.3(a) shows the evolution of the integrator outputs for $X = 0.725$ and $V_r = 1$. The integrator output ranges coincide with the maximum of both curves, that, as shown in Fig. A.3(a), are reached in the beginning ($n = 0$) of a sequence of positive pulses for the first integrator, and in an intermediate point of the sequence for the second integrator. In the latter, the value of the maximum is reached when the derivative of (A.13) equals 0, that is, for

$$n_m = \frac{I_0}{g_1(V_r - X)} - \frac{3V_r + X}{2V_r - X} \qquad (A.16)$$

Thus, the maximum values of both curves are

$$m_1 = I_0 = I_{n_0} + g_1(X + V_r)$$

$$m_2 = J_{n_m} = J_0 + g_2 \frac{[g_1(3V_r + X) - 2I_0]^2}{8g_1(V_r - X)} \qquad (A.17)$$

Until now, the limitation of the maximum level achievable at the integrator output has not been considered. For a first-order modulator, if the integrator OS is smaller than that in (A.8) the modulator does not operate correctly (the averaged output does not coincided with the input level). However, for a second-order modulator, certain integrator OS values, which are smaller than m_2 in (A.17), generate output sequences of such a duration that the maximum value of both integrator outputs are below that OS, so that the operation of the modulator is not degraded. This fact, which is illustrated through behavioral simulation in Fig. A.4, is understood if we take into account, that unlike in first-order modulators, the duration of consecutive pulse sequences is not unique for each input level.

To cope with this difficulty, an iterative procedure is proposed to calculate the integrator OS as a function of the integrator weights and input level. This procedure starts from an arbitrary value of OS and consists of the following steps:

- Consecutive use of expressions (A.16), (A.17), (A.14) and (A.15), taking into account that the maximum value of the integrator output is limited by OS.

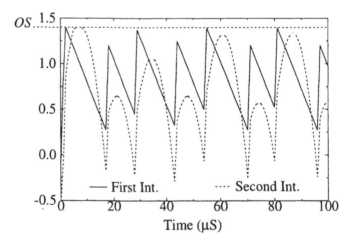

Figure A.4: Effect of the output-swing limitation

- Check if the maximum reached(A.17) during the second and subsequent sequences is larger than OS. This being the case, the value of OS is slightly increased. In general, two or three sequences are enough to check this.

- Repeat the previous process until the maximum reached is below the last value of OS.

Fig. A.5 compares the results obtained using the algorithm above with those obtained through behavioral simulation in the presence of dynamic inputs. A good fitting between both curves is observed. Regarding computational resources, using the proposed algorithm reduces the CPU time by a factor of 100. Note that, for $g_1 = 0.25$, $g_2 = 0.5$ and (X/V_r) close to 0.9, OS should be slightly larger than V_r, while for $g_1 = g_2 = 0.5$, OS must equal at least $2V_r$.

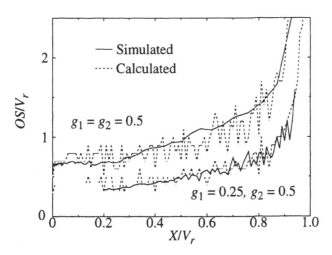

Figure A.5: Output swing required in integrators

Index